Collins

Edexcel GCSE 9-1
Maths
Foundation

Workbook

Trevor Senior

Revision That Really Works

Experts have found that there are two techniques that help you to retain and recall information and consistently produce better results in exams compared to other revision techniques.

It really isn't rocket science either – you simply need to:

- **test yourself** on each topic as many times as possible
- **leave a gap** between the test sessions.

Three Essential Revision Tips

1. Use Your Time Wisely

- Allow yourself plenty of time.
- Try to start revising at least six months before your exams – it's more effective and less stressful.
- Don't waste time re-reading the same information over and over again – it's not effective!

2. Make a Plan

- Identify all the topics you need to revise.
- Plan at least five sessions for each topic.
- One hour should be ample time to test yourself on the key ideas for a topic.
- Spread out the practice sessions for each topic – the optimum time to leave between each session is about one month but, if this isn't possible, just make the gaps as big as you can.

3. Test Yourself

- Methods for testing yourself include: quizzes, practice questions, flashcards, past papers, explaining a topic to someone else, etc.
- Don't worry if you get an answer wrong – provided you check what the correct answer is, you are more likely to get the same or similar questions right in future!

Visit **collins.co.uk/collinsGCSErevision** for more information about the benefits of these techniques, and for further guidance on how to plan ahead and make them work for you.

Command Words used in Exam Questions

This table defines some of the most commonly used command words in GCSE exam questions.

Command word	Meaning
Write...	No working will be required
Write down...	
Find...	A small amount of working will be required
Work out...	Some working will be required
Calculate...	Some working will be required and it is likely a calculator will be needed
Expand...	Remove brackets
Explain...	An explanation (a sentence or mathematical statement) is required
Show...	All working leading to the answer must be shown
Prove...	All steps must be shown and, for a geometrical proof, reasons must be given
Justify...	Show all working or give a written explanation
Simplify fully...	Suggests more than one stage is needed to simplify an expression
Draw...	Suggests accuracy is important
Sketch...	No accurate measurements are needed

Contents

N Number   A Algebra   G Geometry and Measures

S Statistics   P Probability   R Ratio, Proportion and Rates of Change

Number 1, 2 & 3

Grade 1-3 **1** Write 35% as a decimal.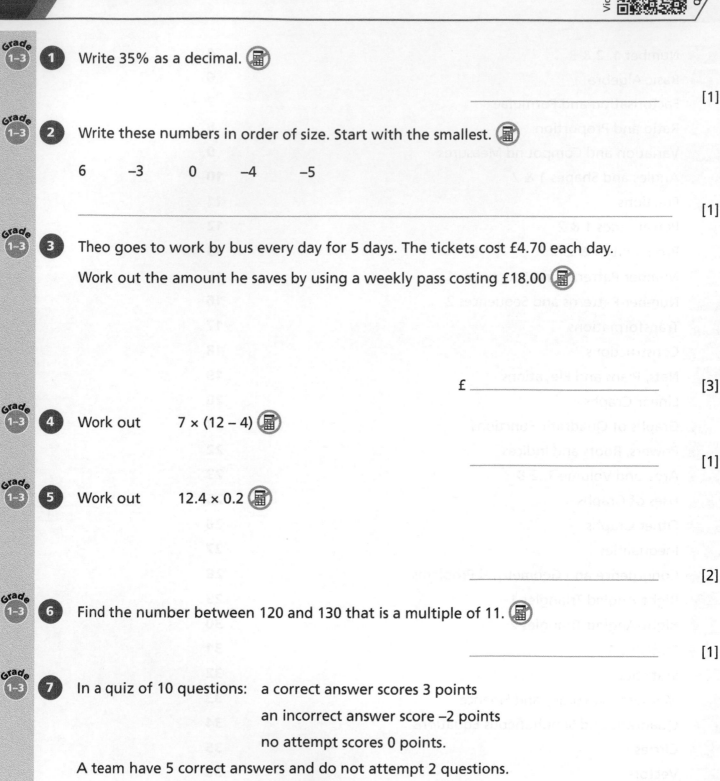

_____ [1]

Grade 1-3 **2** Write these numbers in order of size. Start with the smallest.

6   –3   0   –4   –5

_____ [1]

Grade 1-3 **3** Theo goes to work by bus every day for 5 days. The tickets cost £4.70 each day.

Work out the amount he saves by using a weekly pass costing £18.00

£ _____ [3]

Grade 1-3 **4** Work out   7 × (12 – 4)

_____ [1]

Grade 1-3 **5** Work out   12.4 × 0.2

_____ [2]

Grade 1-3 **6** Find the number between 120 and 130 that is a multiple of 11.

_____ [1]

Grade 1-3 **7** In a quiz of 10 questions: a correct answer scores 3 points

an incorrect answer score –2 points

no attempt scores 0 points.

A team have 5 correct answers and do not attempt 2 questions.

How many points do they score?

_____ [2]

Grade 1-3 **8** Write down two prime numbers that have a sum of 28.

_____ [1]

Grade 1–3

9 Write these values in order of size. Start with the smallest.

$\frac{1}{2}$ 0.2 $\frac{3}{4}$ -1.4 $\frac{1}{12}$

_____ [2]

Grade 3–5

10 a) Write the number 35 million in standard form.

_____ [1]

b) Write 3.49×10^{-3} as an ordinary number.

_____ [1]

Grade 3–5

11 Find the highest common factor (HCF) of 84 and 120.

_____ [3]

Grade 3–5

12 Write 36 as a product of prime factors.

_____ [2]

Grade 3–5

13 Work out the value of $5^3 - 4^2$

_____ [2]

Grade 3–5

14 A green light flashes every 12 seconds. A blue light flashes every 16 seconds.

After how many seconds will both lights flash together?

_____ [3]

Grade 3–5

15 A shop makes sandwiches on white, brown or granary bread.

There are 6 different choices of fillings.

Show that there are 18 different sandwich types.

_____

_____ [1]

Total Marks _____ / 27

Basic Algebra

Grade 1–3 **1** Simplify $3a \times 4b$

_____ [1]

Grade 1–3 **2** Simplify $2x + 8y - x - 3y$

_____ [1]

Grade 1–3 **3** Simplify $w \times w \times w \times w$

_____ [1]

Grade 1–3 **4** $A = 3x + 4y$

Work out the value of A when $x = 5$ and $y = -2$

$A =$ _____ [2]

Grade 1–3 **5** Factorise $9a + 6$

_____ [1]

Grade 3–5 **6** Factorise $8x^2 - 2x$

_____ [2]

Grade 3–5 **7** Expand and simplify $2x(3x + 1) + 5(3x - 2)$

_____ [3]

Grade 3–5 **8** Solve $2x + 4 = x + 1$

$x =$ _____ [2]

Grade 3–5 **9** One side of a square has length $(5x - 2)$ cm.

Work out the perimeter when $x = 3$

$(5x - 2)$ cm

_____ cm [3]

Total Marks _____ / 16

Factorisation and Formulae

Grade 3–5 **1** Expand and simplify $(x + 3)(x - 7)$

_____ [2]

Grade 3–5 **2** Expand and simplify $(4x + 1)(2x - 3)$

_____ [2]

Grade 3–5 **3** Factorise $x^2 + 9x + 18$

_____ [2]

Grade 3–5 **4** Factorise $x^2 - x - 20$

_____ [2]

Grade 3–5 **5** Make x the subject of the formula $y = 3x + 5$

_____ [2]

Grade 3–5 **6** a) Make a the subject of the formula $v^2 = u^2 + 2as$

_____ [2]

b) Work out the value of a when $v = 8$, $u = 5$ and $s = 3$

$a =$ _____ [2]

Grade 3–5 **7** Make A the subject of the formula $r = \sqrt{\dfrac{A}{\pi}}$

_____ [2]

Total Marks _____ / 16

Ratio and Proportion

Grade 1–3

1 Here is a pattern of tiles. Write down the ratio of black tiles to white tiles. 🖩

_____ [1]

Grade 1–3

2 Write down the ratio of £3.50 to 75p. Give your answer in its simplest form. 🖩

_____ [2]

Grade 3–5

3 A woman invests £1650 in shares and cash in the ratio 6 : 5

a) What fraction of the money is in shares?

_____ [1]

b) How much does she invest in shares?

£_____ [2]

Grade 3–5

4 A house is 14.4 metres high. A model is made using the scale 1 : 18

Work out the height of the model. Give your answer in centimetres.

_____ cm [3]

Grade 3–5

5 There are 96 people on an aircraft. Half of them are men. 12 are children.

Work out the ratio in its simplest form   number of women : number of children

_____ [3]

Grade 3–5

6 A tap leaks 45 litres of water over 3 days. How much water will leak in 7 days?

State any assumptions you make.

_____ litres [3]

Total Marks _____ / 15

Variation and Compound Measures

1 An aircraft flew 210 km in 1 hour 30 minutes. Work out the average speed.

_____ km/h   [2]

2 Matt drives at an average speed of 65 mph for 3 hours. How many miles does he drive?

_____   [2]

3 The density of gold is 19.3 g/cm³. A ring has a mass of 4.5 grams and a volume of 0.3 cm³.

Is the ring made of pure gold? Use the formula $\text{Density} = \dfrac{\text{mass}}{\text{volume}}$

_____   [2]

4 A bank has two compound accounts, Saver and Money Maker.

| **Saver**: Interest 1.5% per annum | | **Money Maker**: Interest 2% per annum |

I invest £2000 in Saver and £1500 in Money Maker.

Work out which account will pay more interest after 2 years.

_____   [4]

5 Here is a distance–time graph of a car.

Work out the speed of the car in miles per hour.

_____ mph   [2]

Distance (miles) vs _Time (minutes)_ graph, axes from 0–25 miles and 0–30 minutes.

Total Marks _____ / 12

Angles and Shapes 1 & 2

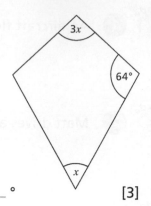

Grade 3–5 **1** The diagram shows a kite.

Work out the size of angle x.

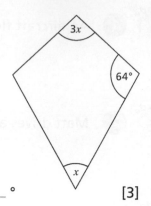

_____ ° [3]

Grade 3–5 **2** The exterior angle of a regular polygon is 18°. Work out the number of sides of the polygon.

_____ [2]

Grade 3–5 **3** The diagram shows an isosceles triangle touching two parallel lines.

Work out the size of angle x.

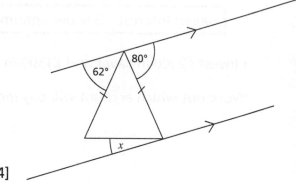

_____ ° [4]

Grade 3–5 **4** **a)** Write down the bearing of Q from P.

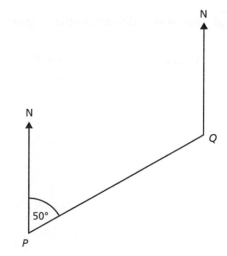

_____ ° [1]

b) Work out the bearing of P from Q.

_____ ° [2]

Total Marks _____ / 12

Fractions

1 Write the fraction $\frac{24}{72}$ in its simplest form.

_____ [1]

2 Find the number that is exactly halfway between $\frac{1}{3}$ and $\frac{5}{6}$

_____ [2]

3 Work out $\frac{3}{4} \times \frac{2}{5}$

Give your answer as a fraction in its simplest form.

_____ [2]

4 Work out $3\frac{1}{2} + 2\frac{2}{3}$

_____ [2]

5 There are 8 people in a class. This is $\frac{2}{5}$ of the class. How many people are in the class altogether?

_____ [3]

6 Which is bigger, $\frac{5}{8}$ of 40 or $\frac{2}{3}$ of 36? You must show your working.

_____ [2]

7 Write $0.\dot{7}$ as a fraction.

_____ [1]

8 180 people attend a party. $\frac{1}{3}$ are men and $\frac{1}{2}$ are women. What fraction are children?

_____ [2]

Total Marks _____ / 15

Percentages 1 & 2

Grade 1–3

1 Write 8% as a decimal.

_____ [1]

Grade 1–3

2 Write 0.24 as a percentage.

_____ % [1]

Grade 1–3

3 Work out   5% of 250 grams

_____ g [2]

Grade 1–3

4 Work out 30 minutes as a percentage of 4 hours.

_____ % [2]

Grade 3–5

5 A coffee machine costs £800. VAT is added at 20%.

The machine is paid for in 12 equal payments. Work out the cost of each payment.

£_____ [3]

Grade 3–5

6 In a sale, the cost of a jacket is reduced from £150 to £105. Work out the percentage reduction.

_____ % [2]

Grade 3–5

7 Jack wants to buy 30 kg of pet food. He sees these two offers:

A	£23.50 for 15 kg bag Offer 10% off		B	£8.50 for 5 kg bag Buy 5 bags get one free

Which is the better offer for Jack, A or B? Show your working.

_____ [4]

Total Marks _____ / 15

Probability 1 & 2

Grade 1-3

1 A four-sided spinner is labelled A, B, C and D.

The table shows the probabilities of the spinner landing on A, on B or on D.

Letter	A	B	C	D
Probability	0.3	0.1		0.4

a) Work out the probability the spinner will land on C.

_____ [1]

b) Give a reason why the spinner is not fair.

_____ [1]

c) The spinner is spun 50 times. Estimate the number of times it lands on A.

_____ [2]

Grade 3-5

2 A bag contains three different shapes: circles, triangles and squares.

The probability of picking a circle at random is 0.1. There are twice as many triangles as squares.

a) Work out the probability of picking a triangle at random.

_____ [2]

b) There are 12 circles in the bag. Work out the number of squares in the bag.

_____ [2]

Grade 3-5

3 There are 7 blue counters and 3 red counters in a bag. A counter is picked at random and replaced.

a) Write down the probability that the counter picked is blue.

_____ [1]

b) A second counter is picked. Work out the probability that both are red.

_____ [2]

4 ξ = {Integers from 11 to 20 inclusive}

A = {prime numbers}   B = {13, 14, 15, 16, 17}

a) Complete the Venn diagram.

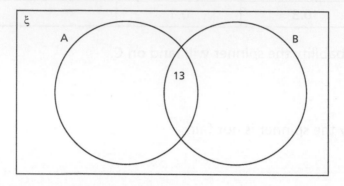

[3]

b) A number is chosen at random from ξ.

Work out the probability that the number is in set A $\cup$ B.

_____ [2]

5 Two boxes, X and Y, each contain 20 counters. Box X has 5 red counters. Box Y has 8 red counters. A counter is taken at random from each box.

a) Complete the tree diagram.

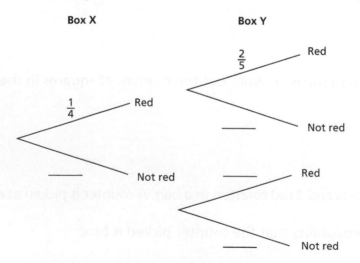

[2]

b) Work out the probability of picking two red counters.

_____ [2]

Total Marks _____ / 20

Number Patterns and Sequences 1

Grade
1–3

1 Here is a pattern of dots.

a) Draw the next pattern in the sequence.

[1]

b) Complete the table for each pattern.

Pattern number	1	2	3	4	5
Number of dots	1	3			

[1]

c) Give one reason why the number of dots does not form an arithmetic sequence.

_____ [1]

2 Here are the first four terms of an arithmetic sequence.   7   10   13   16

a) Write down the next two terms of the sequence.

_____ [1]

b) Is 49 a number in the sequence? Give a reason for your answer.

_____ [2]

3 The nth term of a sequence is $6n + 5$

a) Write down the 10th term in the sequence.

_____ [1]

b) The nth term of a different sequence is $3n - 1$

Find the first number that appears in both sequences.

_____ [2]

4 Work out the missing term of this geometric sequence.

3   6   24   48

_____ [1]

Total Marks _____ / 10

Number Patterns and Sequences 2

1   Here is a pattern of black and white squares:

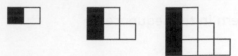

a)   Write down the number of black squares in the 50th pattern.

_____ [1]

b)   Work out the number of white squares in the 4th and 5th patterns.

_____ [2]

2   Here are the first four terms of an arithmetic sequence.   10   14   18   22

Write down an expression, in terms of n, for the nth term of the sequence.

_____ [2]

3   Here are the first four terms of a Fibonacci sequence.   4   4   8   12

a)   Write down the next two terms of the sequence.

_____ [1]

b)   A different Fibonacci sequence starts   0 x x $2x$

Work out the next two terms of the sequence.

_____ [2]

4   For each part, write down the name of the sequence and the nth term.

a)   1   4   9   16   25

_____ [2]

b)   1   8   27   64   125

_____ [2]

Total Marks _____ / 12

Transformations

  1 **a)** Rotate shape **T** 180° about the point (0, 1).

   Label your shape **R**. [2]

   b) Describe fully the single transformation that maps shape **T** onto shape **S**.

   _____

   _____

   _____ [2]

   c) Shape **T** can be transformed to shape **P** using a translation $\begin{pmatrix} a \\ b \end{pmatrix}$

   Write down the values of a and b.

   $a =$ _____ , $b =$ _____ [2]

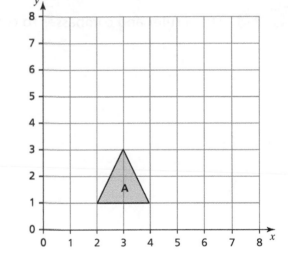

  2 **a)** Reflect shape **A** in the line $y = 4$.

   Label it **B**. [2]

   b) Enlarge shape **A** scale factor 2, centre of enlargement (0, 0).

   Label it **C**. [2]

 3 Describe the **single** transformation that will map triangle **A** to triangle **B**.

   _____

   _____

   _____ [3]

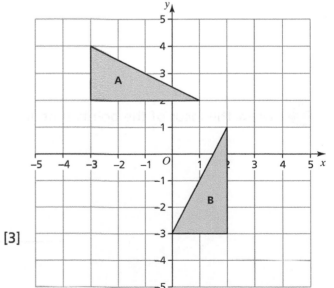

Total Marks _____ / 13

Constructions

Grade 3–5

1 Use a ruler and compasses to construct the perpendicular bisector of the line *AB*.

A ——————— B

[2]

Grade 3–5

2 Use a ruler and compasses to construct the angle bisector of the angle *C*.

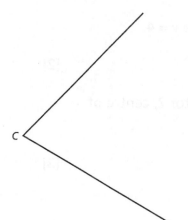

[2]

Grade 3–5

3 Draw the locus of the points that are 2 cm from the line.

———————————————

[2]

Total Marks _____ / 6

Nets, Plans and Elevations

1 Complete the net of the cuboid on the centimetre gird.

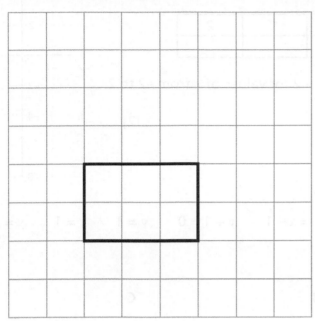

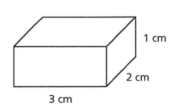

1 cm

2 cm

3 cm

[3]

2 Here is a shape made from cubes:

On the grid, draw the plan view, front elevation and side elevation.

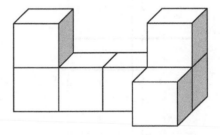

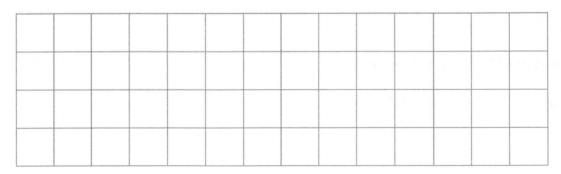

Plan view   Front elevation   Side elevation

[3]

Total Marks _____ / 6

Linear Graphs

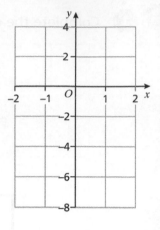

Grade 1–3

1   **a)** Complete the table of values for $y = 3x - 2$ [2]

x	−2	−1	0	1	2
y					

    b) On the grid, draw the graph of $y = 3x - 2$ for values of x from −2 to 2.

      [2]

Grade 3–5

2   Here are six equations. $y = 1 - x$ $y = x - 1$ $y + 1 = 0$ $y = 1$ $x = 1$ $y = x + 1$

Match each graph to its equation.

     A   **B**   **C**

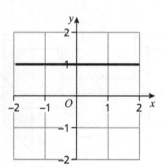   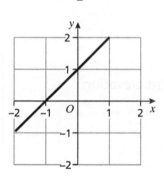   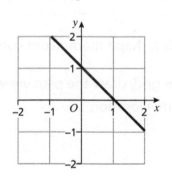

_____   _____   _____   [3]

Grade 3–5

3 A is the point (1, 2). B is the point (3, 8).

   a) Work out the gradient of AB.

       _____   [2]

   b) C is the point (7, 20). Show that ABC is a straight line.

       _____

       _____

       _____   [2]

Total Marks _____ / 11

Graphs of Quadratic Functions

1 a) Complete the table of values for $y = x^2 - x + 2$

x	−2	−1	0	1	2
y			2		4

[2]

b) On the grid, draw the graph of $y = x^2 - x + 2$ for values of x from −2 to 2.

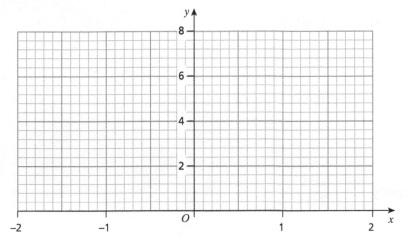

[2]

c) Use your graph to estimate solutions to $x^2 - x + 2 = 3$

$x = $ _____ , $x = $ _____   [2]

2 Here is the graph of $y = x^2 - 2x - 6$

a) Write down the coordinates of the turning point.

(_____ , _____)   [1]

b) Use the graph to find the roots of $x^2 - 2x - 6 = 0$

$x = $ _____ , $x = $ _____   [2]

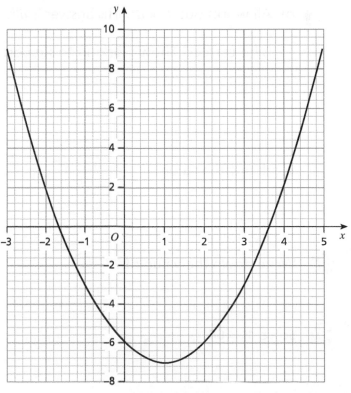

Total Marks _____ / 9

Powers, Roots and Indices

Grade
1–3
1 Write down the cube root of 125.

_____ [1]

Grade
1–3
2 Write 32 as a power of 2.

_____ [1]

Grade
1–3
3 Work out the value of 8^3.

_____ [1]

Grade
3–5
4 **a)** Simplify $(4ab^2)^3$

_____ [2]

b) Simplify $\dfrac{9x^6y^5}{3xy^3}$

_____ [2]

Grade
3–5
5 **a)** Simplify $c^{12} \div c^2$

_____ [1]

b) Ali works out $d^4 \times d^4$. His answer is d^{16}.

Is he correct? Show working to support your answer.

_____ [1]

Grade
3–5
6 Work out the value of $\dfrac{6^8 \times 6^{-4}}{6^3}$

_____ [2]

Grade
3–5
7 Write down the value of 7^{-2}.

_____ [1]

Grade
3–5
8 Solve $4^x = 64$

$x =$ _____ [1]

Total Marks _____ / 13

Area and Volume 1, 2 & 3

1 A right-angled triangle and a rectangle are shown.

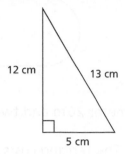 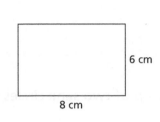

12 cm   13 cm

5 cm

6 cm

8 cm

a) Which shape has the greater area? You must show your working.

_____ [3]

b) Work out the perimeter of the rectangle.

_____ cm [2]

2 This shape is made from two identical rectangles.

Work out the perimeter.

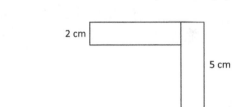

2 cm

5 cm

_____ cm [3]

3 The volume of this cuboid is 72 cm³. Work out the value of x.

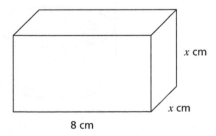

x cm

x cm

8 cm

$x = $ _____ [3]

4 The diagram shows the edge of a race track.

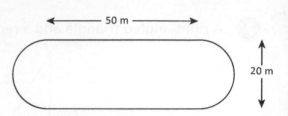

The track is made up of two semi circles with diameter 20 m and two straight pieces.

Fencing is going to be installed around the edge. The fencing costs £24 per metre.

Work out the cost of the fencing. Give your answer to the nearest £10.

£ _____ [5]

5 Work out the area of the trapezium.

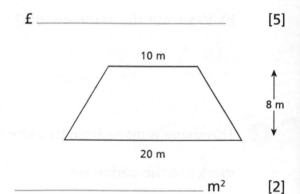

_____ m² [2]

6 1 litre = 1000 cm³

This cylindrical tank is half filled with water.

How much water is in the tank?

Give your answer in litres to 1 decimal place.

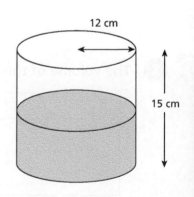

_____ litres [3]

Total Marks _____ / 21

Uses of Graphs

1 A line has equation $y = 3x - 4$

a) What is the gradient of the line?

_____ [1]

b) Write the coordinates of the point where the line intercepts the y-axis.

(_____ , _____) [1]

2 Find the equation of the line that is parallel to the line $y = 5x + 1$ and passes through the point $(0, -3)$.

_____ [2]

3 Show that the lines $y = 2x + 7$ and $2y - 4x - 10 = 0$ are parallel.

_____

_____ [2]

4 You can use this graph to change between kilograms and pounds.

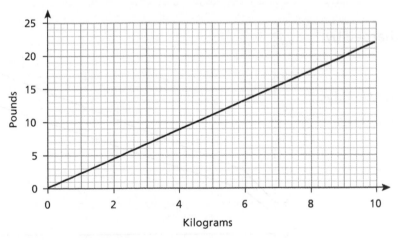

a) Change 5 kilograms to pounds.

_____ pounds [1]

b) Change 60 pounds to kilograms.

_____ kilograms [2]

Total Marks _____ / 9

Other Graphs

Grade 3–5

1 The distance–time graph shows a cycle ride from home to a park and back home.

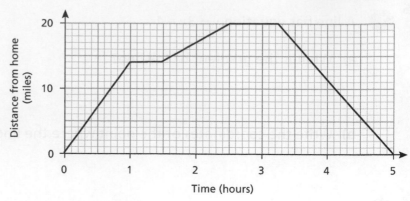

a) How many minutes in duration was the first stop?

_____ [1]

b) What was the average speed on the first part of the ride?

_____ mph   [2]

c) Apart from the stops, which was the slowest section of the ride? Give a reason for your answer.

_____

_____ [1]

Grade 3–5

2 The graph shows two rates charged by a taxi company.

Rate A is from 6 am to 8 pm on Mondays to Fridays. Rate B is for all other times.

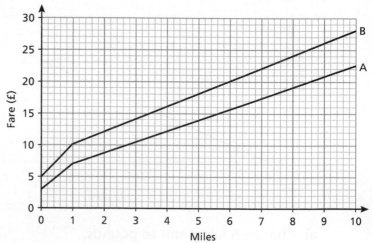

a) How much is the fixed charge for Wednesday at 10 am?

£ _____ [1]

b) How much more is an 8-mile journey at rate B than at rate A?

£ _____ [2]

Total Marks _____ / 7

Inequalities

1 Write one of the symbols =, < or > in each box to make each statement true.

$27 - 8$ ☐ $16 + 4$ 7^2 ☐ 8×6 $3 - 9$ ☐ $4 - 10$ **[2]**

2 On the number line, show the inequality $x > -1$

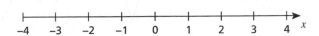

[2]

3 $-1 < x \leqslant 4$ where x is an integer.

Write down all possible values of x.

_____ **[2]**

4 Solve $4x - 7 \geqslant x + 8$

_____ **[2]**

5 Write down the integer values satisfied by this inequality diagram.

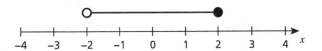

_____ **[2]**

6 $-5 \leqslant x \leqslant 3$

a) Write down the greatest possible value of x^2.

_____ **[1]**

b) Write down the least possible value of x^2.

_____ **[1]**

Total Marks _____ / 12

Congruence and Geometrical Problems

Grade
1–3

1 The diagram shows six triangles on a centimetre grid.

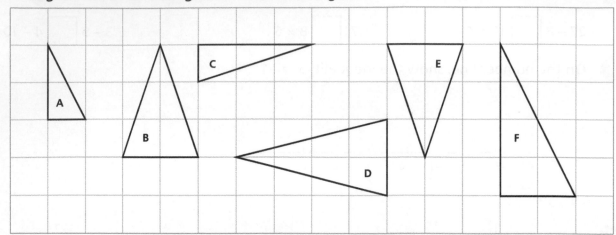

a) Write down the letters of two triangles that are congruent. _____ [1]

b) Write down the letters of two triangles that are similar. _____ [1]

Grade
3–5

2 Here are two similar rectangles:

10 cm

15 cm

x cm

9 cm

Work out the value of *x*.

x = _____ [2]

Grade
3–5

3 Triangle *ABC* and triangle *XYZ* are similar.

a) Work out the length of *BC*.

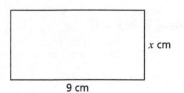

5.8 cm

A 3.4 cm B

Z

3.1 cm

X 1.7 cm Y

_____ cm [2]

b) Work out the perimeter of triangle *XYZ*.

_____ cm [3]

Total Marks _____ / 9

Right-Angled Triangles 1

Grade 3–5

1 *ABC* is a right-angled triangle. Work out the length of *AC*.

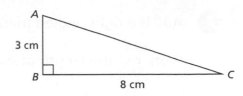

_____ cm   [3]

Grade 3–5

2 *ABC* is a right-angled triangle. Work out the length of *BC*.

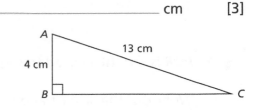

_____ cm   [3]

Grade 3–5

3 The diagram shows two congruent triangles and a square.

Work out the area of the square.

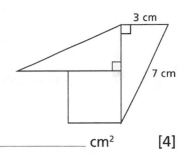

_____ cm²   [4]

Grade 3–5

4 A square field has side 80 metres. Work out the length of a diagonal of the field.

Give your answer to the nearest 10 metres.

_____ m   [3]

Grade 3–5

5 Show that a triangle with sides 5 cm, 12 cm and 13 cm is right-angled.

_____

_____   [2]

Grade 3–5

6 The perpendicular height of this isosceles triangle is 6 cm.

Work out the perimeter.

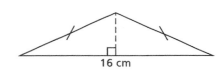

_____ cm   [3]

Total Marks _____ / 18

Right-Angled Triangles 2

Grade 3–5

1 *ABC* is a right-angled triangle.

Work out the length of *AB*.

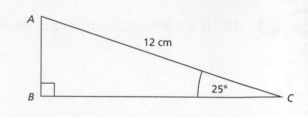

_____ cm   [3]

Grade 3–5

2 *PQR* is a right-angled triangle.

Work out the size of angle *PRQ*.

Give your answer to 1 decimal place.

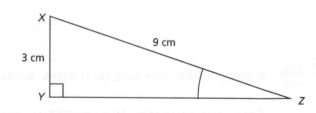

_____ °   [2]

Grade 3–5

3 Here are two right-angled triangles:

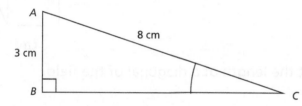

Which angle is bigger: *ACB* or *XZY*? You must give a reason for your answer.

_____

_____   [1]

Grade 3–5

4 The diagram shows a triangular prism.

Work out its volume. Give your answer to 1 decimal place.

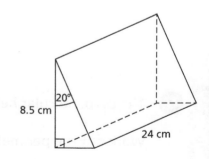

_____ cm³   [4]

Total Marks _____ / 10

Statistics 1

1 The pictogram shows information about the number of coffees served on one day in a café.

Morning	▢ ▢ ▢ ▢
Afternoon	▢ ▢ ▢ ▢ ▢
Evening	

Key: ▢ represents 10 coffees

a) Write down the number of coffees served in the morning. _____ [1]

b) 25 coffees were served in the evening. Show this information on the pictogram. [1]

c) The coffees cost £2.90 each. How much money was taken from the sales of coffee that day?

£ _____ [2]

2 The table shows the number of coins in a jar.

Coin	20p	50p	£1	£2
Number of coins	9	8	3	4

a) Work out the amount of money in the jar.

£ _____ [2]

b) Draw an accurate pie chart to show the information.

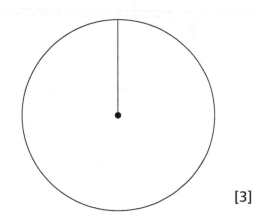

[3]

3 Raj draws this bar chart.

Write down two errors with the chart.

_____

_____

_____

_____ [2]

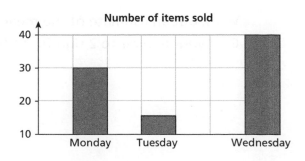

Number of items sold

Monday   Tuesday   Wednesday

Total Marks _____ / 11

Topic-Based Questions

1 Work out the median of these numbers. 7 1 9 5 6 7 2 4

_____ [2]

2 The scatter graph shows information about the marks of some students in two tests.

a) Write down the type of correlation.

_____ [1]

b) One student had a mark that was an outlier.

Circle the cross for that student on the graph. [1]

c) Leo got a mark of 21 on Test A.

Use the scatter graph to estimate his mark on Test B.

_____ [2]

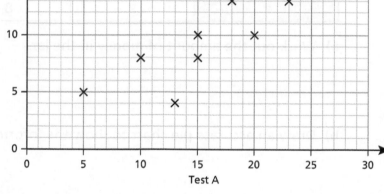

3 The table shows information about the times taken, in minutes, by 15 students to complete a task.

Time (*t* minutes)	Frequency
$0 < t \leqslant 2$	4
$2 < t \leqslant 4$	8
$4 < t \leqslant 6$	3

Work out an estimate of the mean time.
Give your answer to 2 significant figures.

_____ [3]

Total Marks _____ / 9

Measures, Accuracy and Finance

Grade 3–5

1 The length of a pencil is 17.6 cm to 1 decimal place.

Complete the error interval for the pencil.

_____ cm ⩽ length < _____ cm   [2]

Grade 3–5

2 Naby changes £150 into US dollars. The exchange rate is £1 = $1.38

a) Does he have enough dollars to buy jeans costing $70 and a jacket costing $125?

You must show your working.

_____   [3]

b) The same jeans cost £59 in England.

Are the jeans cheaper in England or the USA? You must show your working.

_____   [2]

Grade 3–5

3 A 2000-litre tank is full of water. Water is leaking out at the rate of 15 litres per minute.

a) How many minutes will it take for the tank to half empty?

_____   [2]

b) State one assumption you made in working out your answer to part a).

_____   [1]

Grade 3–5

4 a) Use approximations to estimate the value of 199.4 ÷ 50.2

_____   [2]

b) Is your answer to part a) an underestimate or an overestimate?

Give a reason for your answer.

_____

_____   [1]

Total Marks _____ / 13

Quadratic and Simultaneous Equations

1 Factorise $x^2 - 7x + 12$

_____ [2]

2 $(x + c)(x + 6) = x^2 + dx + 18$

Work out the values of c and d.

$c =$ _____, $d =$ _____ [2]

3 Solve $x^2 - 2x - 15 = 0$

$x =$ _____ or $x =$ _____ [2]

4 The area of this rectangle is $84\,\text{cm}^2$.

x cm

$(x + 5)$ cm

a) Form an equation to show this information.

_____ [1]

b) Solve the equation to work out the value of x.

$x =$ _____ [3]

5 Solve the simultaneous equations $2x + y = 1$

$x - 2y = 8$

$x =$ _____, $y =$ _____ [4]

Total Marks _____ / 14

Circles

9780008326708

Grade 1-3

1 Write down the mathematical names of each of the straight lines on these circles.

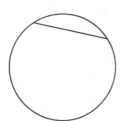   

_____   _____   [2]

Grade 1-3

2 Follow the instruction above each diagram.

Draw a diameter   Draw and shade a segment

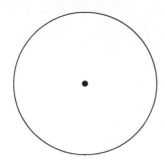   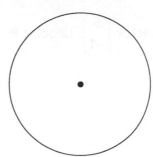

[2]

Grade 3-5

3 Work out the size of angle x.

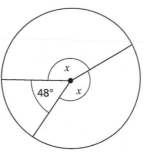

48° x x

_____ °   [2]

Grade 3-5

4 In the diagram O is the centre of the circle.

Work out the size of angle y.

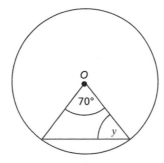

O
70°
y

_____ °   [2]

Total Marks _____ / 8

Vengeance Vectors

Vectors

Grade 3–5

1 $a = \begin{pmatrix} 5 \\ 3 \end{pmatrix}$ $b = \begin{pmatrix} 1 \\ 2 \end{pmatrix}$

Work out $2a - b$ as a column vector.

$\begin{pmatrix} \underline{} \\ \underline{} \end{pmatrix}$ [2]

Grade 3–5

2 $c = \begin{pmatrix} 2 \\ -7 \end{pmatrix}$ $d = \begin{pmatrix} 3 \\ 4 \end{pmatrix}$

Work out $3c + 4d$ as a column vector.

$\begin{pmatrix} \underline{} \\ \underline{} \end{pmatrix}$ [2]

Grade 3–5

3 **a)** Translate shape **A** using the vector $\begin{pmatrix} 5 \\ -1 \end{pmatrix}$. Label it **B**.

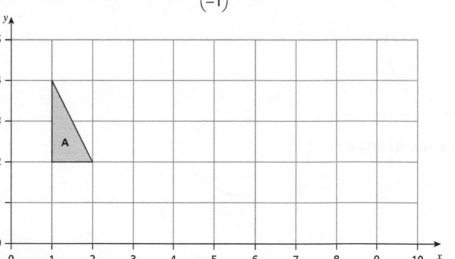

[2]

b) Write down the column vector to translate shape **B** back to shape **A**.

$\begin{pmatrix} \underline{} \\ \underline{} \end{pmatrix}$ [2]

Total Marks _____ / 8

Collins

GCSE
Mathematics
Paper 1 Foundation Tier (Non-Calculator)

F

Time: 1 hour 30 minutes

You must have:

- Ruler graduated in centimetres and millimetres, protractor, pair of compasses, pen, HB pencil, eraser.

You may not use a calculator.

Instructions

- Use **black** ink or black ball-point pen.
- Answer **all** questions.
- Answer the questions in the spaces provided – *there may be more space than you need.*
- **Calculators may not be used.**
- Diagrams are NOT accurately drawn, unless otherwise indicated.
- You must **show all your working out**.

Information

- The total mark for this paper is 80.
- The marks for **each** question are shown in brackets
  - *use this as a guide as to how much time to spend on each question.*
- Read each question carefully before you start to answer it.
- Keep an eye on the time.
- Try to answer every question.
- Check your answers if you have time at the end.

Name: ....

Practice Exam Paper 1

Answer ALL questions.

Write your answers in the spaces provided.

You must write down all stages in your working.

1   Write down the value of 3 in the number 9306

(Total for Question 1 is 1 mark)

2   Write the following numbers in order of size.

       6   –4   0   –3   –1

Start with the smallest.

(Total for Question 2 is 1 mark)

3   Write 0.3 as a fraction.

(Total for Question 3 is 1 mark)

4 Write down a prime number between 30 and 40.

_____

(Total for Question 4 is 1 mark)

5 Write down an even square number that has two digits.

_____

(Total for Question 5 is 1 mark)

6 **(a)** Simplify $\quad 4 \times x \times 3 \times y$

_____ **(1)**

(b) Simplify $\quad 5a + 6b + 7a - b$

_____ **(2)**

(Total for Question 6 is 3 marks)

7 Here are the first four terms of a sequence.

1   8   27   64

(a) Write down the next term of the sequence.

_____ (1)

(b) Write down the name of this sequence.

_____ (1)

(Total for Question 7 is 2 marks)

8 There are 24 coins in a bag.

$\frac{2}{3}$ of the coins are 50p coins.

There is £9.40 altogether.
Six of the coins are 20p coins.

Complete the table for the coins in the bag.

Coin	Number of coins
50p	
20p	6
	Total = 24

(Total for Question 8 is 4 marks)

9   A shopkeeper is looking at who visits her shop one morning.

(a) Complete the tally and frequency column for this information.

Person	Tally	Frequency
Man	||	2
Girl	⊞	
Boy		6
Woman	⊞ ||	

(2)

(b) On the grid, draw a bar chart for this information.

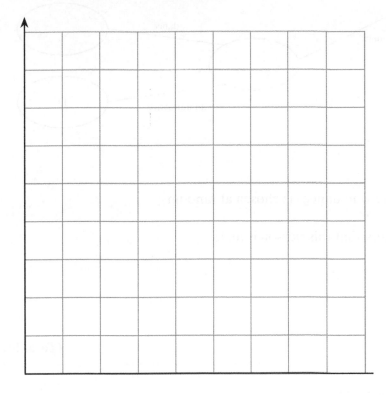

(3)

(Total for Question 9 is 5 marks)

10 120 people go to a party.

75 of these people are female.

84 of the people dance.

13 of the males do not dance.

(a) Use this information to complete the frequency tree.

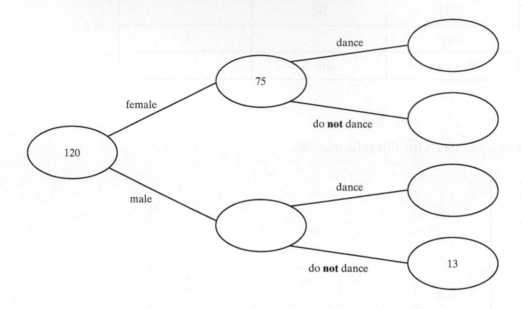

(3)

(b) One of the people who dances is chosen at random.

Find the probability that this person is male.

_____ (2)

(Total for Question 10 is 5 marks)

11 Two paint testers cost £3.50

Work out the cost of five paint testers.

£ _____

(Total for Question 11 is 2 marks)

12 The perimeter of this kite is 19 cm.

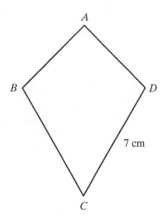

Work out the length of *AB*.

_____ cm

(Total for Question 12 is 2 marks)

13 Here is some information about some students.

	Boys	Girls
Year 7	4	6
Year 8	20	15

(a) Write down the number of Year 7 girls to Year 8 girls as a ratio.
Give your answer in its simplest form.

_____ (1)

(b) What fraction of the boys are in Year 8?
Give your answer in its simplest form.

_____ (2)

(Total for Question 13 is 3 marks)

14 A scale model of a car is created at a scale of 1 : 12
The model has a length of 38 cm.

Work out the actual length of the car.
Give your answer in metres.

_____ m

(Total for Question 14 is 2 marks)

15

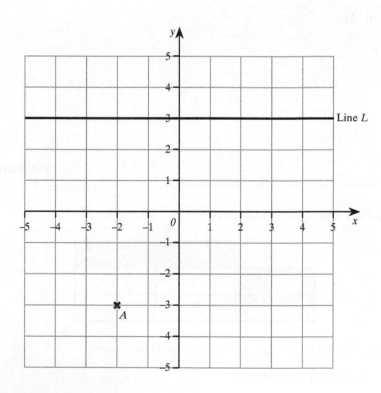

(a) Write down the coordinates of point *A*.

(_____ , _____)   **(1)**

(b) Write down the equation of line *L*.

_____   **(1)**

(c) Work out the shortest distance from point *A* to line *L*.

_____ units   **(1)**

(Total for Question 15 is 3 marks)

16 Sam drives 162 miles from Sheffield to London on the motorway.

50 of the miles have roadworks.

(a) Work out the journey time.

Assume an average speed of: 50 mph through the roadworks
70 mph for the rest of the journey.

_____ (4)

(b) Sam arrives later than expected.

What does this tell you about the average speed?

_____ (1)

(Total for Question 16 is 5 marks)

17 Here is a formula: $V = 3r^2h$

Work out the value of V when $r = 2$ and $h = 6$.

$V = $ _____

(Total for Question 17 is 2 marks)

18   By rounding each number to 1 significant figure, estimate the answer to

$$\frac{139 \times 1.9}{71}$$

(Total for Question 18 is 3 marks)

19   Gino buys three packets of seed for lettuces, carrots and beans from a garden centre.

62% of the lettuce seeds grow. $\frac{3}{5}$ of the carrot seeds grow.

The ratio of bean seeds that grow to the number that do not grow is 2 : 1

Which type has the greatest proportion of seed growing?
You must show your working.

(Total for Question 19 is 4 marks)

20   The diagram shows a regular pentagon made from five identical triangles.

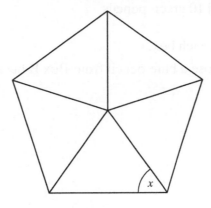

(a) Show that $x = 54°$

(3)

(b) Here is one of the triangles.

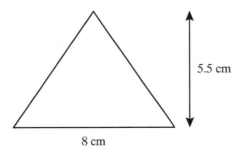

5.5 cm

8 cm

Work out the area of the pentagon.

_____ cm² (4)

(Total for Question 20 is 7 marks)

21   Box A contains 4 blue pencils and 6 green pencils.

Box B contains 6 blue pencils and 10 green pencils.

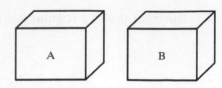

A pencil is taken at random from each box.

Ali says, "I am more likely to choose a blue pencil from Box B because there are more in that box."

Is he correct?

You must show your working.

_____

(Total for Question 21 is 3 marks)

22   Theo invests £500 in a savings account.

It earns simple interest at the rate of 2% per year.

Work out the total value of the savings after three years.

£ _____

(Total for Question 22 is 3 marks)

23   **(a)** Solve $4(x - 5) = 24$

$x =$ _____   (3)

(b) Factorise fully $8x^2 - 12x$

_____   (2)

(Total for Question 23 is 5 marks)

24 **(a)** Write 237 000 in standard form.

_____ (1)

(b) Write 4.5×10^{-4} as an ordinary number.

_____ (1)

(c) Which of the following two numbers is greater?

8.91×10^3 or 5.62×10^4

Give a reason for your answer.

_____ (1)

(Total for Question 24 is 3 marks)

25 In a school, the ratio of boys to girls is $3 : 2$
25% of the boys study French.
50% of the girls study French.

What percentage of the students study French?

_____ %

(Total for Question 25 is 3 marks)

26   Solve $x^2 - 11x + 28 = 0$

(Total for Question 26 is 3 marks)

27    $\mathbf{a} = \begin{pmatrix} 4 \\ 9 \end{pmatrix}$ $\mathbf{b} = \begin{pmatrix} 3 \\ -2 \end{pmatrix}$

(a) Work out 5**a** as a column vector.

 (1)

(b) Work out **a** – 2**b** as a column vector.

$\begin{pmatrix} \\ \end{pmatrix}$ (2)

(Total for Question 27 is 3 marks)

TOTAL FOR PAPER IS 80 MARKS

Collins

GCSE

Mathematics

Paper 2 Foundation Tier (Calculator)

F

Time: 1 hour 30 minutes

You must have:

- Ruler graduated in centimetres and millimetres, protractor, pair of compasses, pen, HB pencil, eraser, calculator

Instructions

- Use **black** ink or black ball-point pen.
- Answer **all** questions.
- Answer the questions in the spaces provided – *there may be more space than you need.*
- **Calculators may be used.**
- If your calculator does not have a π button, take the value of π to be 3.142 unless the question instructs otherwise.
- Diagrams are NOT accurately drawn, unless otherwise indicated.
- You must **show all your working out**.

Information

- The total mark for this paper is 80.
- The marks for each question are shown in brackets
  - *use this as a guide as to how much time to spend on each question.*
- Read each question carefully before you start to answer it.
- Keep an eye on the time.
- Try to answer every question.
- Check your answers if you have time at the end.

Name: _____

Practice Exam Paper 2

Answer ALL questions.

Write your answers in the spaces provided.

You must write down all stages in your working.

1   Write $\frac{3}{20}$ as a percentage.

_____ %

(Total for Question 1 is 1 mark)

2   Write 5.37 correct to 1 decimal place.

_____

(Total for Question 2 is 1 mark)

3   Work out the cube root of 27.

_____

(Total for Question 3 is 1 mark)

4   **(a)** Change 4.16 kilograms to grams.

_____ g   **(1)**

(b) Change 350 centimetres to metres.

_____ m   **(1)**

(Total for Question 4 is 2 marks)

5 Here are four numbers:

$\frac{1}{3}$ $\frac{3}{10}$ 0.33   0.325

Write the numbers in order of size.
Start with the smallest.

_____

(Total for Question 5 is 2 marks)

6 Clare visits a bookshop and buys two books.
Each book costs £4.99
There is an offer of two books for £7.50
She thinks using the offer she will save more than £2.

Is she correct?
You must show your working.

(Total for Question 6 is 3 marks)

7   The diagram shows a triangle.

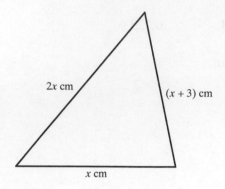

2x cm   (x + 3) cm

x cm

(a) Work out an expression, in terms of x, for the perimeter.

Simplify your answer.

_____ (2)

(b) Write down the type of triangle if $x = 3$

You must show your working.

_____ (3)

(Total for Question 7 is 5 marks)

8   Here is a number machine.

input ⟶ ×7 ⟶ −3 ⟶ output

(a) Work out the output when the input is 5.

_____ (1)

(b) Work out the input when the output is 18.

_____ (1)

(c) Here is a different number machine.

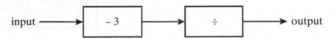

input ⟶ −3 ⟶ ÷ ⟶ output

When the input is 20 the output is 8.5

Complete the number machine.

(1)

(Total for Question 8 is 3 marks)

9   There are 32 counters in a bag.
16 are red, 9 are yellow and the rest are green.
One of the counters is chosen at random.

Work out the probability that the counter is green.

_____

(Total for Question 9 is 2 marks)

10 $T = 7x + 8y$

Work out the value of T when $x = 6$ and $y = -2$

$T =$ _____

(Total for Question 10 is 2 marks)

11   The diagram shows a circular pond in a square garden.
The pond covers 30% of the garden.

Work out the area remaining.

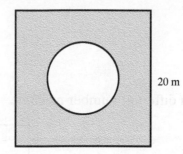

20 m

_____ m²

(Total for Question 11 is 3 marks)

12   Write down the integer values of x where $4 < x \leqslant 9$

_____

(Total for Question 12 is 2 marks)

13 The diagram shows five identical squares on a grid.

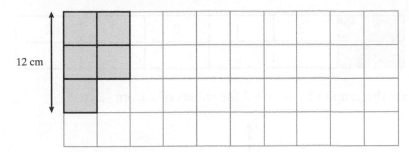

Work out the width and length of the grid.

Width _____ cm

Length _____ cm

(Total for Question 13 is 4 marks)

14 The pictogram shows information about the number of plants in a garden in the four seasons.

Season	
Spring	◯◯◯◯
Summer	◯◯◯◯◯◯
Autumn	◯◯◖
Winter	◯◺

Key:
◯ represents 8 plants

(a) Write down the modal season.

_____ **(1)**

(b) How many plants were there in winter?

_____ **(1)**

(c) Write the number of plants in spring to the number of plants in summer as a ratio.
Give your answer in its simplest form.

_____ **(2)**

(Total for Question 14 is 4 marks)

15 **(a)** Complete the table of values for $y = 2x + 3$ **(2)**

x	-2	-1	0	1	2
y					

(b) On the grid, draw the graph of $y = 2x + 3$ for values of x from -2 to 2.

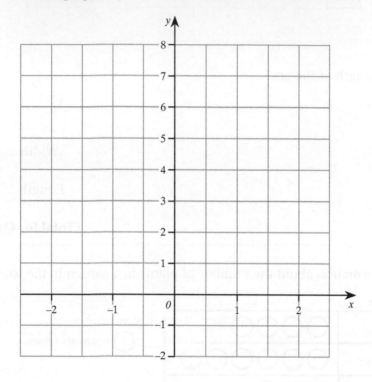

(2)

(c) Solve $2x + 3 = 2$

$x =$ _____ **(2)**

(Total for Question 15 is 6 marks)

16 **(a)** Here is a list of numbers. 8 5 2 6 3 3

Work out the median.

_____ (2)

(b) Here is a different list of numbers. 7 x $3x$ 6 9
The mean is 40.

Work out the value of x.

$x = $ _____ (3)

(Total for Question 16 is 5 marks)

17 Work out 150% of 82

_____

(Total for Question 17 is 2 marks)

18 The diagram shows a quarter-circle with radius 3.5 cm.

Work out the area.

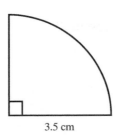

3.5 cm

_____ cm^2

(Total for Question 18 is 3 marks)

19 A group of 72 students take part in tasting some biscuits labelled A, B and C.
The pie chart shows information about the results.
Twice as many chose Biscuit B as their favourite than Biscuit A.

Work out the number of students who chose Biscuit C
as their favourite.

Favourite Biscuit

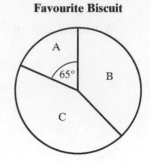

(Total for Question 19 is 4 marks)

20 **(a)** The density of copper is 9 grams/cm³
The volume of copper used to make a bracelet is 2.5 cm³

Work out the mass of the bracelet.

_____ g   **(2)**

(b) A force of 50 Newtons acts on an area of 16 cm².
The force stays the same but the area increases.

What happens to the pressure?

$$\text{pressure} = \frac{\text{force}}{\text{area}}$$

_____ **(1)**

(Total for Question 20 is 3 marks)

21 **(a)** Find the highest common factor (HCF) of 36 and 54.

_____ (2)

(b) Find the lowest common multiple (LCM) of 9, 12 and 18.

_____ (2)

(Total for Question 21 is 4 marks)

22 Describe fully the single transformation that maps shape **A** onto shape **B**.

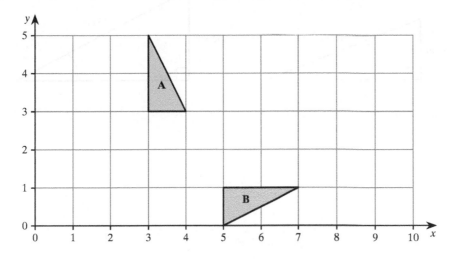

_____

_____

_____

(Total for Question 22 is 3 marks)

23 **(a)** Work out $\dfrac{(839 \times 74)}{\sqrt{97}}$

Write down all the figures on your calculator display.

_____ (2)

(b) Write your answer to part (a) correct to 3 significant figures.

_____ (1)

(Total for Question 23 is 3 marks)

24 These two triangles are similar.

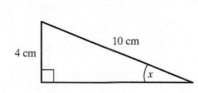

Work out the value of _y_.

$y = $ _____

(Total for Question 24 is 2 marks)

25 The results when an ordinary, six-sided dice is rolled 60 times are shown.

Number shown	1	2	3	4	5	6
Frequency	5	8	12	15	9	11

(a) Do you think the dice is fair?

Give a reason for your answer.

_____

_____ (1)

(b) Work out the relative frequency of rolling a 5 or a 6.

_____ (2)

(Total for Question 25 is 3 marks)

26 A supermarket carries out an online survey of 50 customers.
15 customers say that they usually visit the supermarket on Fridays.

(a) On one Friday, approximately 4500 customers visit the supermarket.

Use this information to estimate how many customers the supermarket has altogether.

_____ (2)

(b) State any assumptions you made.

_____

_____ (1)

(Total for Question 26 is 3 marks)

27 Here are two piles of exercise books.
Each book is 0.75 cm thick.
The smaller pile is 11.25 cm high.

The height of the smaller pile is $\frac{3}{5}$ of the height of the taller pile.

Work out the total number of books in the two piles.

(Total for Question 27 is 4 marks)

TOTAL FOR PAPER IS 80 MARKS

Collins

GCSE

Mathematics

Paper 3 Foundation Tier (Calculator)

F

Time: 1 hour 30 minutes

You must have:

- Ruler graduated in centimetres and millimetres, protractor, pair of compasses, pen, HB pencil, eraser, calculator.

Instructions

- Use **black** ink or black ball-point pen.
- Answer **all** questions.
- Answer the questions in the spaces provided – *there may be more space than you need.*
- **Calculators may be used.**
- If your calculator does not have a π button, take the value of π to be 3.142 unless the question instructs otherwise.
- Diagrams are NOT accurately drawn, unless otherwise indicated.
- You must **show all your working out**.

Information

- The total mark for this paper is 80.
- The marks for **each** question are shown in brackets
  - *use this as a guide as to how much time to spend on each question.*
- Read each question carefully before you start to answer it.
- Keep an eye on the time.
- Try to answer every question.
- Check your answers if you have time at the end.

Name: ....

Practice Exam Paper 3

Answer ALL questions.

Write your answers in the spaces provided.

You must write down all stages in your working.

1 Write down a multiple of 6 between 20 and 40.

_____

(Total for Question 1 is 1 mark)

2 Work out $\sqrt{11.56}$

_____

(Total for Question 2 is 1 mark)

3 Write 3% as a fraction.

_____

(Total for Question 3 is 1 mark)

4 Work out $1 - \dfrac{5}{12}$

_____

(Total for Question 4 is 1 mark)

5 Work out $\dfrac{2}{5}$ of 100

_____

(Total for Question 5 is 2 marks)

6   An ordinary fair dice is thrown once.
The diagram shows a probability scale.

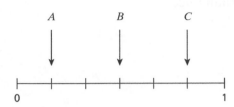

(a) Which letter shows the probability of the dice landing on a number less than 6?

_____ **(1)**

(b) Which letter shows the probability of the dice landing on an even number?

_____ **(1)**

(Total for Question 6 is 2 marks)

7   Here is a 3-D shape.

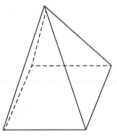

(a) Write down the name of this shape.

_____ **(1)**

(b) Write down the number of faces on this shape.

_____ **(1)**

(c) Bilal says, "The number of faces plus the number of vertices is 2 more than the number of edges."

Is he correct?
You must show your working.

_____

_____ **(2)**

(Total for Question 7 is 4 marks)

8   Make a list of all the possible two-digit numbers that can be made using the digits 7, 8 or 9.

You can use each digit more than once.

(Total for Question 8 is 2 marks)

9   A hotel charges: £75 per night for a room Monday to Friday inclusive

£85 per night for a room Saturday or Sunday

Breakfast each day is £8.70 extra per person

Evening meal each day is £19.50 extra per person.

Dan books a room for Friday night, Saturday night and Sunday night.

He has breakfast on one day only.

He has an evening meal on two days.

Work out the amount he pays altogether.

£ _____

(Total for Question 9 is 4 marks)

10 You are given that 10 inches = 25 cm

(a) Use the information given to complete a conversion graph.

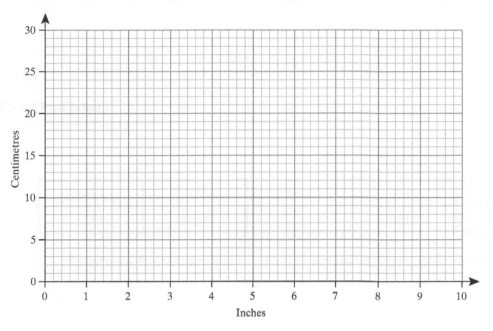

(2)

(b) Change 5 inches to centimetres.

_____ cm   (1)

(c) Change 75 centimetres to inches.

_____ in   (2)

(Total for Question 10 is 5 marks)

11 Solve $4x - 7 = 21$

$x =$ _____

(Total for Question 11 is 2 marks)

12   Factorise fully $12x + 18y + 24z$

(Total for Question 12 is 2 marks)

13   Here is part of a bus timetable:

Sunnyside	1118	1148	1218	1248
Rotherham	1152	1222	1252	1322

(a) How many minutes does the journey take from Sunnyside to Rotherham?

(2)

(b) Trevor wants to arrive in Rotherham before noon.
It takes him 15 minutes to walk to the bus stop in Sunnyside.

What is the latest time he should set off to catch his bus?

(2)

(Total for Question 13 is 4 marks)

14   Increase 250 by 12%

(2)

(Total for Question 14 is 2 marks)

15 **(a)** Expand $2x(x-4)$

_____ **(2)**

(b) Simplify $y \times y^2 \times y^3$

_____ **(1)**

(Total for Question 15 is 3 marks)

16 **(a)** The diagram shows a parallelogram $VQRW$ inside a rectangle $PQST$.

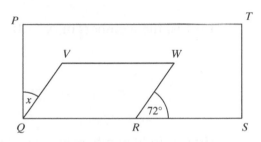

Work out the size of angle x.

_____° **(2)**

(b) The diagram shows an equilateral triangle and an isosceles triangle.

Work out the size of angle ADC.
Give a reason for each stage of your working.

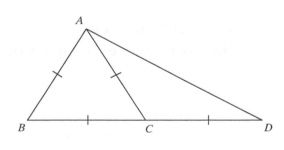

_____° **(3)**

(Total for Question 16 is 5 marks)

17 Mo runs at 7.5 metres per second.

Carl runs at 25 kilometres per hour.

Who runs faster?
You must show your working.

_____

(Total for Question 17 is 3 marks)

18 A = {even numbers between 5 and 15}

B = {factors of 24}

(a) List the members of A ∩ B

_____ **(2)**

(b) C = {odd numbers less than 20}

Describe the set A ∩ C

_____ **(1)**

(Total for Question 18 is 3 marks)

19 *ABC* is a right-angled triangle.

Work out the length *BC*.
Give your answer correct to 1 decimal place.

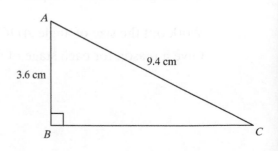

_____ cm

(Total for Question 19 is 3 marks)

20 A shop sells tins of dog food in packs of 6 and packs of 4.

1 pack of 6   £3.80   1 pack of 4   £2.35
Buy 2 packs and get 20% off   2 packs for   £3.95

What is the cheapest way to buy 12 tins?
You must show your working.

(Total for Question 20 is 4 marks)

21 Work out the value of $\dfrac{5^4 \times 5^6}{5^8 \times 5^{-1}}$

(Total for Question 21 is 3 marks)

22 Three teams A, B and C play a game.

The probability that B wins to the probability that C wins is in the ratio 2 : 3

The probability that C wins = 0.54

Work out the probability that A wins.

(Total for Question 22 is 4 marks)

---

23 The times taken by 60 people to get a vaccine are shown.

Time (t minutes)	$0 < t \leqslant 5$	$5 < t \leqslant 10$	$10 < t \leqslant 15$	$15 < t \leqslant 20$
Frequency	22	14	13	11

(a) Find the interval that contains the median.

_____ **(1)**

(b) On the grid, draw a frequency polygon for the information in the table.

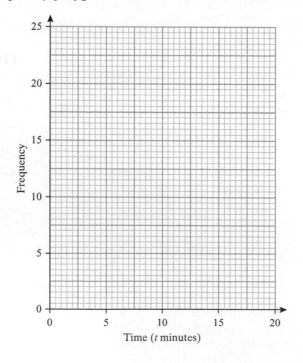

(2)

(Total for Question 23 is 3 marks)

24 1 gallon = 4.5 litres

The diagram shows a large empty water tank.

It holds 250 gallons when full.

Water is pumped into the tank at the rate of 12 litres per minute.

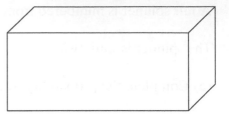

How long will it take to fill the tank?

Give your answer in hours and minutes to the nearest minute.

(Total for Question 24 is 3 marks)

25 A fair spinner is numbered from 1 to 5.

The spinner is spun twice.

(a) Complete the probability tree diagram to show the probabilities of landing on even or odd numbers.

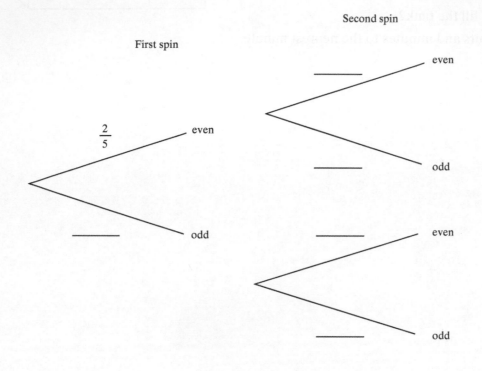

First spin

Second spin

$\frac{2}{5}$ even

even

odd

_____ odd

even

odd

(2)

(b) Work out the probability of both spins landing on an even number.

_____ (2)

(Total for Question 25 is 4 marks)

26   Solve the simultaneous equations $3x + 2y = 11$

$x - 4y = 13$

$x =$ _____

$y =$ _____

(Total for Question 26 is 3 marks)

27 **(a)** Complete the table of values for $y = x^2 - 5x + 2$

x	–1	0	1	2	3	4	5	6
y		2			–4		2	

(2)

(b) On the grid, draw the graph of $y = x^2 - 5x + 2$ for values of x from –1 to 6.

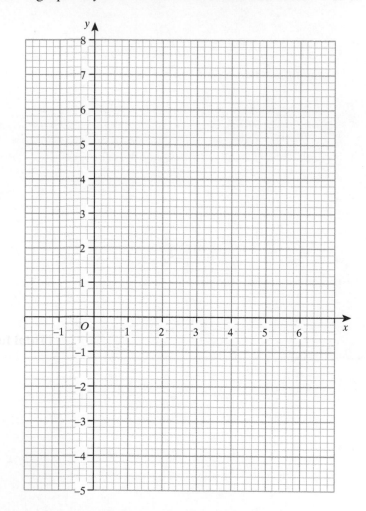

(2)

(c) Use your graph to estimate the coordinates of the turning point.

(_____ , _____) (2)

(Total for Question 27 is 6 marks)

TOTAL FOR PAPER IS 80 MARKS

Answers

Workbook Answers

You are encouraged to show all your working out, as you may be awarded marks for method even if your final answer is wrong. Full marks can be awarded where a correct answer is given without working but, if a question asks for working, you must show it to gain full marks. If you use a correct method that is not shown in the answers below, you would still gain full credit for it.

Page 4: Number 1, 2 & 3

1. 0.35 [1]
2. −5, −4, −3, 0, 6 [1]
3. $5 \times £4.70 = £23.50$ [1]
   £23.50 − £18 = £5.50 [1]
4. $7 \times 8 = 56$ [1]
5. $124 \times 2 = 248$ [1]
   $248 \div 100 = 2.48$ [1]
6. 121 [1]
7. $5 \times 3 = 15$ and $3 \times -2 = -6$ [1]
   $15 − 6 = 9$ [1]
8. 5 and 23 or 11 and 17 [1]
9. $-1.4, \frac{1}{12}, 0.2, \frac{1}{2}, \frac{3}{4}$ [2]
   [1 mark for any three in correct order]
10. a) 3.5×10^7 [1]
    b) 0.003 49 [1]
11. $84 = 2^2 \times 3 \times 7$ [1]
    $120 = 2^3 \times 3 \times 5$ [1]
    $HCF = 2^2 \times 3 = 12$ [1]
12. 2, 2, 3, 3 [1]
    $2 \times 2 \times 3 \times 3$ or $2^2 \times 3^2$ [1]
13. $5^3 = 125$ [1]
    $125 − 16 = 109$ [1]
14. Multiples of 12: 12, 24, 36, 48, … [1]
    Multiples of 16: 16, 32, 48, … [1]
    48 seconds [1]
15. $3 \times 6 (= 18)$ [1]

Page 6: Basic Algebra

1. $12ab$ [1]
2. $x + 5y$ [1]
3. w^4 [1]
4. $3 \times 5 + 4 \times -2$ or $15 − 8$ [1]
   7 [1]
5. $3(3a + 2)$ [1]
6. $2x(4x − 1)$ [2]   [1 mark for $2x(4x…)$]
7. $6x^2 + 2x + 15x − 10$ [2]
   [1 mark for $6x^2 + 2x$; 1 mark for $+ 15x − 10$]
   $6x^2 + 17x − 10$ [1]
8. $2x − x = 1 − 4$ [1], $x = −3$ [1]
9. $5 \times 3 − 2 = 13$ [1]
   Perimeter is 4×13 cm [1]
   52 cm [1]

Page 7: Factorisation and Formulae

1. $x^2 − 7x + 3x − 21$ [1]
   $x^2 − 4x − 21$ [1]
2. $8x^2 − 12x + 2x − 3$ [1]
   $8x^2 − 10x − 3$ [1]
3. $(x + 3)(x + 6)$ [2]
   [1 mark for $(x + a)(x + b)$ where $ab = 18$]
4. $(x + 4)(x − 5)$ [2]
   [1 mark for $(x + a)(x + b)$ where $ab = −20$]
5. $y − 5 = 3x$ [1]
   $x = \dfrac{y − 5}{3}$ [1]
6. a) $v^2 − u^2 = 2as$ [1]
   $a = \dfrac{v^2 − u^2}{2s}$ [1]
   b) $8^2 = 5^2 + 2(a)3$ or $8^2 − 5^2 = 2(a)3$ [1]
   $a = \dfrac{13}{2}$ or 6.5 [1]
7. $r^2 = \dfrac{A}{\pi}$ [1]
   $A = \pi r^2$ [1]

Page 8: Ratio and Proportion

1. 4 : 12 or 1 : 3 [1]
2. 350 : 75 [1]
   14 : 3 [1]
3. a) $\dfrac{6}{11}$ [1]
   b) $\dfrac{6}{11} \times 1650$ or 1 part = £150 [1]
   £900 [1]
4. $14.4 \div 18$ or $1440 \div 18$ [1]
   80 cm [1]
5. $96 \div 2 = 48$ (men) [1]
   $96 − 48 − 12 = 36$ (women) [1]
   36 : 12 = 3 : 1 [1]
6. $45 \div 3 = 15$ (litres per day) [1]
   $15 \times 7 = 105$ litres [1]
   Assume same rate [1]

Page 9: Variation and Compound Measures

1. $210 \div 1.5$ [1] = 140 km/h [1]

   > Average speed = Distance ÷ Time

2. 65×3 [1] = 195 (miles) [1]
3. $4.5 \div 0.3$ or 19.3×0.3 or $4.5 \div 19.3$ [1]
   15 or 5.79 or 0.23… and No [1]
4. $2000 \times \left(1 + \dfrac{1.5}{100}\right)^2$ or $1500 \times \left(1 + \dfrac{2}{100}\right)^2$ [1]
   2060.45 [1]
   1560.60 [1]
   £60.45 and £60.60 and Money Maker [1]
5. $25 \div 0.5$ [1]
   50 mph [1]

Page 10: Angles and Shapes 1 & 2

1. $360 - 64 - 64 = 232$ **[1]**
    $232 \div 4$ **[1]** $= 58°$ **[1]**
2. $360 \div 18$ **[1]** $= 20$ **[1]**
3. $180 - 80 - 62 = 38$ **[1]**
    $(180 - 38) \div 2$ **[1]** $= 71$ **[1]**
    $80 - 71 = 9°$ **[1]**
4. a) 050° **[1]**
    b) $180 + 50$ **[1]** $= 230°$ **[1]**

Page 11: Fractions

1. $\frac{1}{3}$ **[1]**
2. $\frac{1}{3} = \frac{4}{12}$ and $\frac{5}{6} = \frac{10}{12}$ or $\frac{1}{3} + \frac{5}{6} = \frac{7}{6}$ **[1]**
    $\frac{7}{12}$ **[1]**
3. $\frac{6}{20}$ **[1]**; $\frac{3}{10}$ **[1]**
4. $3 + \frac{3}{6} + 2 + \frac{4}{6}$ or $5 + \frac{7}{6}$ **[1]**
    $6\frac{1}{6}$ **[1]**
5. $\frac{1}{5}$ is $8 \div 2 = 4$ (people) **[1]**
    4×5 **[1]** $= 20$ **[1]**
6. $\frac{5}{8} \times 40 = 25$ or $\frac{2}{3} \times 36 = 24$ **[1]**
    25 and 24 and $\frac{5}{8}$ of 40 identified **[1]**
7. $\frac{7}{9}$ **[1]**
8. $\frac{1}{3}$ of $180 = 60$ and $\frac{1}{2}$ of $180 = 90$
    or $1 - \frac{1}{3} - \frac{1}{2}$ **[1]**
    $\frac{30}{180}$ or $\frac{1}{6}$ **[1]**

Page 12: Percentages 1 & 2

1. 0.08 **[1]**
2. 24% **[1]**
3. 0.05×250 **[1]** $= 12.5$ **[1]**
4. $\frac{30}{240} \times 100$ (%) **[1]** $= 12.5\%$ **[1]**
5. $800 \times 1.2 = £960$ **[1]**
    $960 \div 12$ **[1]** $= £80$ **[1]**
6. $\frac{45}{150} \times 100$ (%) **[1]** $= 30\%$ **[1]**
7. $£23.50 \times 2 = £47$ **[1]**
    $£47 \times 0.9$ **[1]**
    $£8.50 \times 5$ **[1]**
    £42.30 and £42.50 and A cheaper. **[1]**

Page 13: Probability 1 & 2

1. a) 0.2 **[1]**
    b) Different probability for each letter **[1]**
    c) 0.3×50 **[1]** $= 15$ **[1]**

Remember that exhaustive mutually exclusive probabilities add up to 1.

2. a) $(1 - 0.1) \div 3$ or 0.3 **[1]**
        $0.3 \times 2 = 0.6$ **[1]**
    b) 12×3 **[1]** $= 36$ **[1]**
3. a) $\frac{7}{10}$ **[1]**
    b) $\frac{3}{10} \times \frac{3}{10}$ **[1]** $= \frac{9}{100}$ **[1]**
4. a)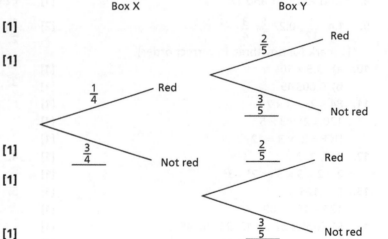
        [3]
        [1 mark for 11 and 19 in correct place;
        1 mark for 14, 15 and 16 in correct place]
    b) $\frac{7}{10}$ **[2]**
        [1 mark for identifying correct region]
5. a)
        [2]
        [1 mark for any correct branch]
    b) $\frac{1}{4} \times \frac{2}{5}$ **[1]** $= \frac{1}{10}$ **[1]**

Page 15: Number Patterns and Sequences 1

1. a) **[1]**

    b)

Pattern number	1	2	3	4	5
Number of dots	1	3	6	10	15

        [1]

    c) Term-to-term rule is not adding on the
        same number each time (+2, +3, +4 …) **[1]**

2. a) 19, 22 [1]

 b) Yes, it is the 15th term in the sequence ....
46, 49, 52, ... [1]

3. a) $6 \times 10 + 5 = 65$ [1]

 b) 1st sequence: 11, 17, 23, 29, 35, ... or
2nd sequence 2, 5, 8, 11, 14, 17, ... [1]
11 [1]

4. 12 [1]

Page 16: Number Patterns and Sequences 2

1. a) 50 [1]

 b) $6 + 4 = 10$ **[1]**, $10 + 5 = 15$ **[1]**

2. $4n + 6$ [2]
[1 mark for $4n + ...$]

3. a) $8 + 12 = 20$ [1]
$12 + 20 = 32$ [1]

 b) $x + 2x = 3x$ [1]
$2x + 3x = 5x$ [1]

> For a Fibonacci sequence, add two terms together
> to get the next term.

4. a) Square numbers **[1]**, n^2 **[1]**

 b) Cube numbers **[1]**, n^3 **[1]**

Page 17: Transformations

1. a) Fully correct [2]
[1 mark for any 180° rotation of shape **T**]

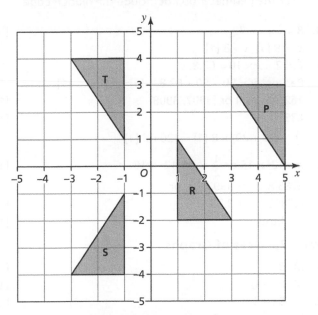

 b) Reflection **[1]** in the x-axis **[1]**

 c) $a = 6$ **[1]**, $b = -1$ **[1]**

> For a translation sort out the x direction first, then
> the y direction.

2. a) Fully correct [2]
[1 mark for any reflection in a line $y = c$]

 b) Fully correct [2]
[1 mark for any enlargement scale factor 2]

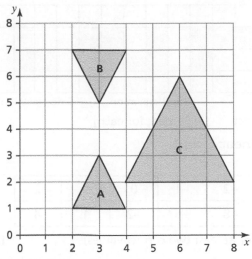

3. Rotation **[1]** by 90° anticlockwise **[1]** about (2, 2) **[1]**

Page 18: Constructions

1.

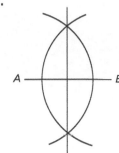

Fully correct [2]
[1 mark for equal intersecting
arcs from A and B]

2.

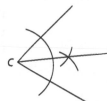

Fully correct [2]
[1 mark for correct arcs]

3.

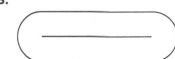

Fully correct [2]
[1 mark for straight lines 2 cm from given line or semi-
circular arcs radius 2 cm]

Page 19: Nets, Plans and Elevations

1.

Fully correct [3]
[1 mark for two correct faces; 2 marks for five correct
faces]

2.

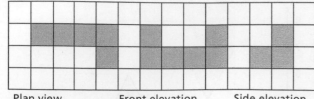

Plan view   Front elevation   Side elevation

Fully correct   **[3]** [1 mark for each]

Page 20: Linear Graphs

1. a)

x	−2	−1	0	1	2
y	−8	−5	−2	1	4

[1 mark for at least two correct values]   **[2]**

b)

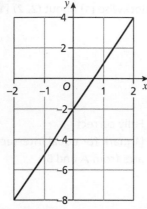

Fully correct   **[2]**
[1 mark for plotting at least two points correctly]

2. A: $y = 1$ **[1]**
B: $y = x + 1$ **[1]**
C: $y = 1 - x$ **[1]**

3. a) $\frac{8 - 2}{3 - 1}$ **[1]** $= 3$ **[1]**

b) $\frac{20 - 2}{7 - 1}$ or $\frac{20 - 8}{7 - 3}$ **[1]**

3 same gradient and common point
so straight line   **[1]**

Page 21: Graphs of Quadratic Functions

1. a)

x	−2	−1	0	1	2
y	8	4	2	2	4

[1 mark for at least two correct values]   **[2]**

b)

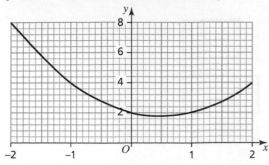

[1 mark for points correctly plotted; 1 mark
for points joined with smooth curve]   **[2]**

c) $x = -0.6$ **[1]**, $x = 1.6$ **[1]**

2. a) $(1, -7)$ **[1]**
b) $x = -1.6$ **[1]**, $x = 3.6$ **[1]**

Page 22: Powers, Roots and Indices

1. 5   **[1]**
2. 2^5 **[1]**
3. 512   **[1]**
4. a) $64a^3b^6$ **[2]**
[1 mark for two correct terms]
b) $3x^5y^2$ **[2]**
[1 mark for two correct terms]
5. a) c^{10} **[1]**
b) Not correct, answer is d^8 **[1]**
6. $\frac{6^4}{6^3}$ **[1]** $= 6$ **[1]**

7. $\frac{1}{7^2} = \frac{1}{49}$ **[1]**

8. $x = 3$ **[1]**

Page 23: Area and Volume 1, 2 & 3

1. a) Area of triangle is $\frac{1}{2} \times 5 \times 12$ **[1]**
$= 30$ cm²   **[1]**
Area of rectangle is $8 \times 6 = 48$ cm² and chooses
rectangle   **[1]**
b) $(8 + 6) \times 2$ **[1]** $= 28$ cm **[1]**
2. $5 + 2 + (5 + 2) + 5 + 2 + (5 - 2)$ **[1]** $= 24$ cm **[1]**

> For the perimeter do **not** include the hidden edge.

3. $8 \times x \times x = 72$ **[1]**
$x^2 = 9$ **[1]**, $x = 3$ **[1]**
4. $C = 2 \times \pi \times 10 = 62.8...$ **[1]**
Perimeter $= 50 + 50 + 62.8...$ **[1]** $= 162.8...$ **[1]**
$162.8... \times 24$ or [3907, 3908]   **[1]**
£3910   **[1]**
5. $\frac{1}{2} \times (10 + 20) \times 8$ **[1]** $= 120$ m² **[1]**
6. $\pi \times 12^2 \times 15$ or $\frac{1}{2} \times \pi \times 12^2 \times 15$ **[1]**
6785.8... or 3392.9... or
6.7858... or 3.3929...   **[1]**
3.4 litres   **[1]**

Page 25: Uses of Graphs

1. a) 3   **[1]**
b) $(0, -4)$ **[1]**
2. $y = 5x - 3$ **[2]**
[1 mark for $y = 5x ...$ or gradient $= 5$]
3. $y = 2x + 7$ has gradient 2   **[1]**
$2y - 4x - 10 = 0 \rightarrow y = 2x + 5$; also gradient
2 so parallel   **[1]**
4. a) 11 pounds   **[1]**
b) For example, 10 pounds $= 4.5$ kg   **[1]**
60 pounds is 6×4.5 kg $= 27$ kg   **[1]**

> Read off at 10 pounds, then scale up.

Page 26: Other Graphs

1. a) 30 minutes   **[1]**
b) 14 miles (in 1 hour) **[1]** so 14 mph **[1]**

c) 3rd section (from 1.5 to 2.5 hours) **[1]**
    Least slope (lowest gradient) **[1]**
2. a) £3 **[1]**
    b) £24 − £19 **[1]** = £5 **[1]**

Page 27: Inequalities

1. 27 − 8 (19) < 16 + 4 (20)
    7² (49) > 8 × 6 (48)
    3 − 9 (−6) = 4 − 10 (−6) **[2]**
    [1 mark if two correct]
2.

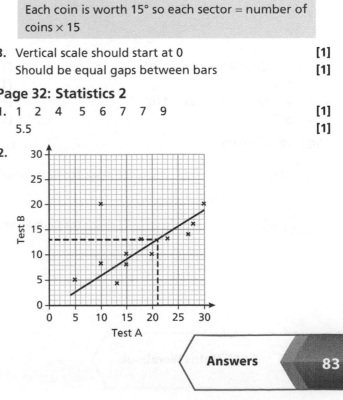

    $\overset{\circ}{\underset{-4 \quad -3 \quad -2 \quad -1 \quad 0 \quad 1 \quad 2 \quad 3 \quad 4}{\longmapsto}}$ x **[2]**

    [1 mark for a filled in circle plus arrow or open circle at
    −1 without the arrow]
3. 0, 1, 2, 3, 4 **[2]**
    [1 mark for four correct or five correct with one extra
    incorrect]
4. $4x - x \geqslant 8 + 7$ or $3x \geqslant 15$ **[1]**
    $x \geqslant 5$ **[1]**
5. −1, 0, 1, 2 **[2]**
    [1 mark for three correct or four correct
    with one incorrect]
6. a) 25 **[1]**
    b) 0 **[1]**

> If you square a negative number, you get a
> positive number.

Page 28: Congruence and Geometrical Problems

1. a) B and E **[1]**
    b) A and F **[1]**
2. Scale factor 1.5 or $\frac{2}{3}$ or $\frac{15}{9}$ or $\frac{5}{3}$ or $\frac{9}{15}$ or $\frac{3}{5}$ **[1]**

    $9 \div 1.5$ or $9 \times \frac{2}{3}$ or $10 \div \frac{15}{9}$ or $10 \times \frac{9}{15} = 6$ **[1]**
3. a) Scale factor 2 or 0.5 or 3.1 cm × 2 **[1]**
        6.2 cm **[1]**
    b) 5.8 ÷ 2 = 2.9 or 5.8 + 3.4 + 6.2 = 15.4 **[1]**
        15.4 ÷ 2 or 2.9 + 1.7 + 3.1 **[1]** = 7.7 cm **[1]**

Page 29: Right-Angled Triangles 1

1. 3² + 8² or 9 + 64 **[1]**
    $\sqrt{9 + 64}$ **[1]** $\sqrt{73}$ or 8.5... cm **[1]**
2. 13² − 4² or 169 − 16 **[1]**
    $\sqrt{169 - 16}$ **[1]** $\sqrt{153}$ or 12.36... cm **[1]**
3. 7² − 3² or 49 − 9 **[1]**
    $\sqrt{49 - 9}$ **[1]** $\sqrt{40}$ or 6.32... cm
    Side of square = 3.32... cm **[1]**
    Area = (3.32...)² = 11 cm² **[1]**
4. 80² + 80² or 6400 + 6400 **[1]**
    $\sqrt{6400 + 6400}$ or $\sqrt{12800}$ or 113. (...) **[1]**
    110 m **[1]**
5. 5² + 12² or 25 + 144 **[1]**
    169 = 13² **[1]**
6. 6² + 8² or 36 + 64 **[1]**
    $\sqrt{36 + 64}$ or $\sqrt{100}$ or 10 **[1]**
    10 + 10 + 16 = 36 cm **[1]**

Page 30: Right-Angled Triangles 2

1. $\sin 25° = \frac{AB}{12}$ or $AB = 12 \sin 25°$ **[1]**
    5.07 cm **[1]**
2. $\cos PRQ = \frac{5.9}{8.6}$ or $\cos PRQ = 0.686...$ **[1]**
    46.7° **[1]**
3. ACB is bigger as sin ACB = 0.375 and sin XZY = 0.333...
    or angle ACB = 22° and angle XZY = 19.4...° **[1]**
4. $\tan 20° = \frac{h}{8.5}$ or $h = 8.5 \tan 20°$ **[1]**
    h = 3.09... **[1]**
    Volume = $\frac{1}{2} \times 8.5 \times 3.09... \times 24$ **[1]**
    = 315.6 cm³ **[1]**

Page 31: Statistics 1

1. a) 40 **[1]**
    b)

Morning	☐ ☐ ☐ ☐
Afternoon	☐ ☐ ☐ ☐ ☐
Evening	☐ ☐ ☐
[1]

    c) 40 + 45 + 25 = 110 **[1]**
        110 × £2.90 = £319 **[1]**
2. a) 20p × 9 + 50p × 8 + £1 × 3 + £2 × 4 = **[1]**
        £1.80 + £4 + £3 + £8
        £16.80 **[1]**
    b) 360° ÷ 24 = 15°; 9 × 15 = 135° **[1]**
        8 × 15° = 120°, 3 × 15° = 45°, 4 × 15° = 60° **[1]**

[1]

> Each coin is worth 15° so each sector = number of
> coins × 15

3. Vertical scale should start at 0 **[1]**
    Should be equal gaps between bars **[1]**

Page 32: Statistics 2

1. 1 2 4 5 6 7 7 9 **[1]**
    5.5 **[1]**
2.

a) Positive [1]

b) (10, 20) identified [1]

c) Line of best fit drawn [1]

   Correct reading from graph, e.g. 12 or 13 [1]

3. Uses mid-class values 1, 3, 5 [1]

   $((1 \times 4) + (3 \times 8) + (5 \times 3)) \div 15$ [1]

   $(4 + 24 + 15) \div 15 = 2.86\ldots = 2.9$ [1]

Page 33: Measures, Accuracy and Finance

1. 17.55 [1], 17.65 [1]

2. a) 150×1.38 [1] = $207 [1]

   $70 + $125 = $195 and enough [1]

   b) $70 \div 1.38$ or 59×1.38 [1]

   £50.72 or $81.42 and cheaper in USA [1]

3. a) $1000 \div 15$ [1] = 66.6 minutes or 67 minutes [1]

   b) Rate stays the same [1]

4. a) $200 \div 50$ [1] = 4 [1]

   b) Overestimate as numerator increased and denominator decreased [1]

Page 34: Quadratic and Simultaneous Equations

1. $(x - 3)(x - 4)$ [2]

   [1 mark for $(x - a)(x - b)$ where $ab = 12$]

2. $c = 3$ [1], $d = 9$ [1]

3. $(x - 5)(x + 3)$ [1], $x = 5$ or $x = -3$ [1]

4. a) $x(x + 5) = 84$ [1]

   b) $x^2 + 5x = 84$ or $x^2 + 5x - 84 = 0$ [1]

   $(x - 7)(x + 12)(= 0)$ [1], $x = 7$ [1]

5. $4x + 2y = 2$ or $2x - 4y = 16$ [1]

   $5x = 10$ or $5y = -15$ [1]

   $x = 2$ or $y = -3$ [1], $x = 2$ and $y = -3$ [1]

Page 35: Circles

1. chord [1] tangent [1]

2. Fully correct [2]   [1 mark for one suitable answer]

   Draw a diameter   Draw and shade a segment

   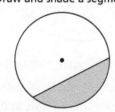

3. $x + x + 48° = 360°$ or $2x = 312°$ [1]

   $x = 156°$ [1]

4. $y + y + 70° = 180°$ or $2y = 110°$ [1]

   $y = 55°$ [1]

Page 36: Vectors [2]

1. $\begin{pmatrix} 9 \\ 4 \end{pmatrix}$ [2]   [1 mark for each value]

2. $\begin{pmatrix} 18 \\ -5 \end{pmatrix}$ [2]   [1 mark for each value]

3. a)

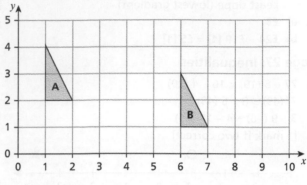

b) $\begin{pmatrix} -5 \\ 1 \end{pmatrix}$ [2]   [1 mark for each value]

Pages 37–50: Practice Exam Paper 1

1. 300 or three (3) hundred(s) [1]

2. −4, −3, −1, 0, 6 [1]

3. $\frac{3}{10}$ [1]

4. 31 or 37 [1]

5. 16 or 36 or 64 [1]

6. (a) $12xy$ [1]

   (b) $12a + 5b$ [1 mark for one correct term] [2]

7. (a) 125 [1]

   (b) Cube numbers [1]

8.

Coin	Number of coins
50p	16
20p	6
10p	2
	Total = 24

$\frac{2}{3} \times 24$ [1] = 16 (50p coins) [1]

$24 - 16 - 6 = 2$ (coins left) [1]

$16 \times 50p = £8$, $6 \times 20p = £1.20$, so 20p left

over giving $2 \times 10p$ [1]

9. (a)

Person	Tally	Frequency
Man	||	2
Girl	++++	5
Boy	++++ |	6
Woman	++++ ||	7

[2]

[1 mark for each column]

(b) Fully correct   **[3]**

[1 mark for vertical scale; 1 mark for labels;
1 mark for bars of correct height equally spaced]

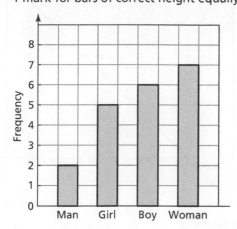

10. (a)

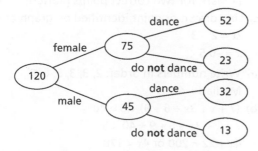

Fully correct **[3]** [1 mark for 45; 1 mark for 32 and 52]

You will need to work out the number of people who do not dance.

(b) $\frac{32}{84}$ or $\frac{8}{21}$ **[2]** [1 mark for 32 identified]

11. £3.50 ÷ 2 or £1.75   **[1]**
£1.75 × 5 = £8.75   **[1]**

12. (19 − 2 × 7) or 5   **[1]**
(19 − 2 × 7) ÷ 2 or 5 ÷ 2 = 2.5 cm   **[1]**

Remember the properties of a kite.
$BC = BD$ and $AB = AD$.

13. (a) 6 : 15 = 2 : 5   **[1]**

(b) $\frac{20}{24}$ **[1]** $= \frac{5}{6}$ **[1]**

14. 38 × 12 = 456 cm   **[1]**
456 ÷ 100 = 4.56 m   **[1]**

15. (a) (−2, −3)   **[1]**
(b) $y = 3$ **[1]**
(c) 6 units   **[1]**

16. (a) 162 − 50 = 112 miles at 70 mph   **[1]**
112 ÷ 70 or 11.2 ÷ 7 or 50 ÷ 50 or 1   **[1]**
1.6   **[1]**
1.6 + 1 = 2.6 hours or 2 hours 36 minutes   **[1]**

(b) Average speed was less than assumed   **[1]**

17. 3 × 2² × 6 or 3 × 2 × 2 × 6 **[1]** = 72 **[1]**

18. $\frac{140 \times 2}{70}$ **[2]** [1 mark for two of 140, 2 and 70]

4   **[1]**

It is easier to cancel down before multiplying.

19. $\frac{3}{5}$ = 60%   **[1]**
2 : 1 means 2 parts out of 3   **[1]**
2 parts out of 3 is two-thirds or 66.6% or 67%   **[1]**
Beans   **[1]**

20. (a) Each angle at centre = 360° ÷ 5 = 72°   **[1]**
In a triangle this leaves 180° − 72° = 108°   **[1]**
Triangle is isosceles so 108° ÷ 2 = 54°   **[1]**

(b) Area of one triangle = $\frac{1}{2} \times 8 \times 5.5$ **[1]**
= 22 cm² **[1]**
22 × 5 **[1]** = 110 cm² **[1]**

21. Box A P(blue) = $\frac{4}{10}$ or $\frac{2}{5}$ or 0.4   **[1]**

Box B P(blue) = $\frac{6}{16}$ or $\frac{3}{8}$ or 0.375   **[1]**

Not correct as 0.4 > 0.375   **[1]**

22. Interest = £ $\frac{500 \times 2 \times 3}{100}$ **[1]**
Interest = £30   **[1]**
Total value = £500 + £30 = £530   **[1]**

Remember that simple interest is the same amount for each year.

23. (a) 4x − 20 = 24   **[1]**
4x = 24 + 20 or 4x = 44   **[1]**
x = 11   **[1]**
(b) 4x(2x − 3)   **[2]**
[1 mark for partial factorisation, e.g. 2x(4x − 6) or x(8x − 12)]

Check that you have taken out all the factors to factorise fully.

24. (a) 2.37 × 10⁵   **[1]**
(b) 0.000 45   **[1]**
(c) 5.62 × 10⁴ is greater as a larger power of 10   **[1]**

25. Assume, for example, 100 students in the school
so 60 boys and 40 girls   **[1]**
60 ÷ 4 = 15 boys study French and 20 girls study French   **[1]**
15 + 20 = 35% study French   **[1]**

26. (x − 4)(x − 7)   **[2]**
[1 mark for (x − a)(x − b) where ab = 28]
x = 4 or x = 7   **[1]**

You will need to factorise into two brackets first.

27. (a) $\begin{pmatrix} 20 \\ 45 \end{pmatrix}$ **[1]**

(b) $\begin{pmatrix} 4 \\ 9 \end{pmatrix} + -2 \begin{pmatrix} 3 \\ -2 \end{pmatrix} = \begin{pmatrix} 4 \\ 9 \end{pmatrix} - \begin{pmatrix} 6 \\ -4 \end{pmatrix} = \begin{pmatrix} -2 \\ 13 \end{pmatrix}$ **[2]**

[1 mark for each part]

Pages 51–64: Practice Exam Paper 2

1. 15% [1]
2. 5.4 [1]
3. 3 [1]
4. (a) 4160 g [1]

   > Remember that 1 kg = 1000 g

   (b) 3.5 m [1]

   > Remember that 1 m = 100 cm

5. $\frac{3}{10}$, 0.325, 0.33, $\frac{1}{3}$ [2]

   [1 mark for any three in correct order]

   > To compare, change all values to the same form, e.g. decimals.

6. 2 × £4.99 = £9.98 [1]
   £9.98 − £7.50 [1]
   £2.48 and correct [1]
7. (a) $2x + x + x + 3$ [1] $= (4x + 3)$ cm [1]
   (b) $2x = 6$ [1]
   $x + 3 = 6$ [1]
   So 3, 6, 6 is isosceles [1]
8. (a) 32 [1]
   (b) 3 [1]
   (c)

   input → −3 → ÷2 → output [1]

9. 32 − 16 − 8 = 7 green [1]

   $\frac{7}{32}$ [1]

10. $7 × 6 + 8 × -2$ or $42 − 16$ [1] = 26 [1]
11. $20 × 20 = 400$ [1]

   $\frac{30}{100} × 400$ or 120 or $\frac{70}{100} × 400$ [1] = 280 m² [1]

12. 5, 6, 7, 8, 9 [2] [1 mark for four correct and no more than one incorrect]

   > Integers are whole numbers.

13. Each square has side 12 ÷ 3 = 4 cm [1]
    Width is 4 × 4 [1], length is 10 × 4 [1]
    Width = 16 cm and Length = 40 cm [1]

    > Work out the width of one square first.

14. (a) Summer [1]
    (b) 8 + 2 = 10 [1]
    (c) 32 : 48 or 4 : 6 [1] 2 : 3 [1]
15. (a)

x	−2	−1	0	1	2
y	−1	1	3	5	7

    [2] [1 mark for three or four correct values]

(b)

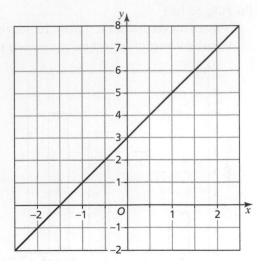

Fully correct [2]
[1 mark for two correct points plotted]
(c) Line drawn or point identified on graph at $y = 2$ or
$2x = 2 − 3$ [1]
$x = −0.5$ [1]

16. (a) Write numbers in order: 2, 3, 3, 5, 6, 8 [1]
    Median = 4 [1]
    (b) $(7 + x + 3x + 6 + 9) ÷ 5 = 40$
    $7 + x + 3x + 6 + 9 = 40 × 5$ [1]
    $4x + 22 = 200$ or $4x = 178$ [1]
    $x = 44.5$ [1]

    > If the mean of five numbers is 40, then they add up to 40 × 5 = 200.

17. $\frac{150}{100} × 82$ [1] = 123 [1]
18. Area of circle is $\pi × 3.5^2$ or 38.48... cm² [1]
    Area of quarter-circle = $0.25 × \pi × 3.5^2$ or
    $0.25 × 38.48...$ [1]
    9.62 cm² or 9.6 cm² [1]
19. $65° × 2 = 130°$ [1]
    $360° − 65° − 130° = 165°$ [1]

    $\frac{165}{360} × 72°$ [1] = 33 [1]

20. (a) Mass is $2.5 × 9$ [1] = 22.5 g [1]

    > Mass = Volume × Density

    (b) Pressure decreases [1]
21. (a) Factors of 36: 1, 2, 3, 4, 6, 9, 12, 18, 36
    or $36 = 2^2 × 3^2$
    Factors of 54: 1, 2, 3, 6, 9, 18, 27, 54 or
    $54 = 2 × 3^3$ [1]
    HCF = $2 × 3^2 = 18$ [1]
    (b) Multiples of 9: 9, 18, 27, 36, ... or $9 = 3^2$
    Multiples of 12: 12, 24, 36, 48, ... or
    $12 = 2^2 × 3$
    Multiples of 18: 18, 36, 54, 72, ... or
    $18 = 2 × 3^2$ [1]
    LCM = $2^2 × 3^2 = 36$ [1]
22. Rotation [1] 90° clockwise [1] about (3, 1) [1]
23. (a) 6303.878 201 [2] $\left(1 \text{ mark for } \frac{62086}{9.8488...}\right)$
    (b) 6300 [1]

24. $4 : 10 = y : 15$ or $\frac{4}{10} = \frac{y}{15}$ **[1]**

$y = 6$ **[1]**

25. **(a)** Not fair as frequencies vary too much **[1]**

(b) $9 + 11 = 20$ **[1]**

$\frac{20}{60}$ or $\frac{1}{3}$ **[1]**

26. **(a)** 15 out 50 or $\frac{15}{50}$ is 4500 customers so 30% or

$4500 \div 30 \times 100$ **[1]** $= 15\,000$ **[1]**

(b) Assume the sample is representative of all customers **[1]**

27. $11.25 \div 0.75 = 15$ books (smaller pile) **[1]**

$15 \div 3 \times 5$ **[1]** $= 25$ books (larger pile) **[1]**

40 books altogether **[1]**

Pages 65–78: Practice Exam Paper 3

1. 24 or 30 or 36 **[1]**

2. 3.4 **[1]**

3. $\frac{3}{100}$ **[1]**

4. $\frac{7}{12}$ **[1]**

5. $\frac{2}{5} \times 100$ **[1]** $= 40$ **[1]**

6. **(a)** C **[1]**

(b) B **[1]**

7. **(a)** Square-based pyramid **[1]**

(b) 5 **[1]**

(c) 5 edges and 8 vertices seen **[1]**

$5 + 5 = 10$ and $10 - 8 = 2$ and correct **[1]**

8. 77, 78, 79, 87, 88, 89, 97, 98, 99 **[2]**

[1 mark for at least seven correct]

9. £75 + £85 + £85 = £245 **[1]**

£19.50 × 2 = £39 **[1]**

£245 + £8.70 + £39 **[1]** = £292.70 **[1]**

10. (a)

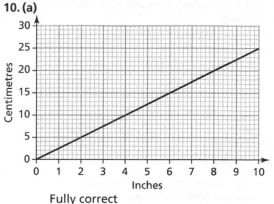

Fully correct **[2]**

[1 mark for two correct points plotted, e.g. (0, 0) and (10, 25)]

(b) 12.5 cm **[1]**

(c) 30 in **[2]**

[1 mark for any scaling method, e.g. 25 cm × 3]

> Use the conversion for 25 cm and scale up by a factor of 3.

11. $4x = 21 + 7$ or $4x = 28$ **[1]**

$x = 7$ **[1]**

12. $6(2x + 3y + 4z)$ **[2]** [1 mark for a partial factorisation, e.g. $3(4x + 6y + 8z)$]

13. **(a)** 52 − 18 **[1]** = 34 **[1]**

(b) Chooses the first bus on the timetable, e.g. 1118 − 15 minutes **[1]** = 1103 **[1]**

14. 250 × 1.12 **[1]** = 280 **[1]**

15. **(a)** $2x^2 - 8x$ **[2]** [1 mark for each term]

(b) y^6 Add the powers. **[1]**

16. **(a)** Angle $VQS = 72°$ (corresponding angles) **[1]**

$x = 90° - 72°$, $x = 18°$ **[1]**

(b) Angle $ACB = 60°$

(angle in an equilateral triangle) **[1]**

Angle $ACD = 180° - 60° = 120°$

(angles on a straight line add up to 180°) **[1]**

Angle $ADC = (180° - 120°) \div 2 = 30°$

(base angles in isosceles triangle are equal) **[1]**

17. 7.5 metres per second = 7.5 ÷ 1000 kilometres per second or 25 km/h = 25 × 1000 metres per hour **[1]**

= 7.5 ÷ 1000 × 60 × 60 kilometres per hour or 25 × 1000 ÷ (60 × 60) metres per second **[1]**

27 km/h or 6.9m/s and Mo runs faster **[1]**

18. **(a)** A ∩ B = {6, 8, 12} **[2]**

[1 mark for two correct and none incorrect]

(b) Empty set, i.e. no members **[1]**

19. $3.6^2 + BC^2 = 9.4^2$ or $BC^2 = 9.4^2 - 3.6^2$ **[1]**

$BC^2 = \sqrt{9.4^2 - 3.6^2} = 8.68\ldots$ **[1]**

8.7 cm **[1]**

20. Two packs of 6 cost £3.80 × 2 = £7.60 **[1]**

With 20% off, cost £7.60 × 0.8 = £6.08 **[1]**

Packs of 4 cost £2.35 + £3.95 **[1]** = £6.30

so two packs of 6 cheaper. **[1]**

21. 5^3 **[2]** [1 mark for each part of $\frac{5^{10}}{5^7}$]

$= 125$ **[1]**

22. P(B wins) = 0.54 ÷ 3 × 2 **[1]** = 0.36 **[1]**

P(A wins) = 1 − 0.54 − 0.36 **[1]** = 0.1 **[1]**

23. **(a)** $5 < t \leqslant 10$ **[1]**

(b)

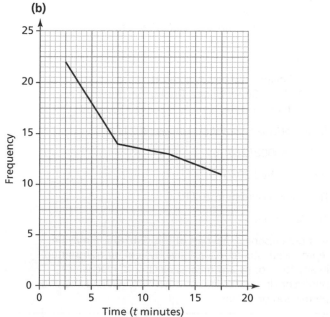

Fully correct [1 mark for correct points used] **[2]**

24. Number of litres = 250 × 4.5 = 1125   **[1]**
1125 ÷ 12 = 93.75 minutes   **[1]**
1 hour 34 minutes   **[1]**

25. (a)

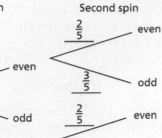

First spin   Second spin

[2] [1 mark for one correct odd branch]

(b) $\frac{2}{5} \times \frac{2}{5}$ **[1]** $= \frac{4}{25}$ **[1]**

26. $6x + 4y = 22$ or $3x - 12y = 39$ **[1]**
$7x = 35$ or $14y = -28$ or $x = 5$ or $y = -2$ **[1]**
$x = 5$ and $y = -2$ **[1]**

> To match coefficients, either multiply the first equation by 2 on both sides or multiply the second equation by 3 on both sides.

27. (a)

x	−1	0	1	2	3	4	5	6
y	8	2	−2	−4	−4	−2	2	8

Fully correct   **[2]**
[1 mark for at least three correct values]

(b)

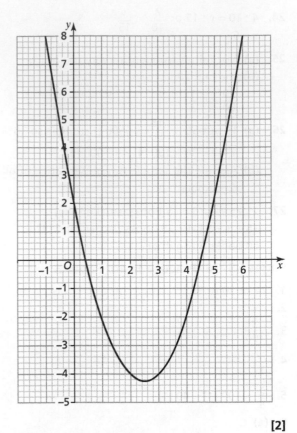

[2]

(c) (2.5, −4.25) **[2]** [accept −4.2 or −4.3 for y coordinate; 1 mark for each coordinate]

ACKNOWLEDGEMENTS

Every effort has been made to trace copyright holders and obtain their permission for the use of copyright material. The author and publisher will gladly receive information enabling them to rectify any error or omission in subsequent editions. All facts are correct at time of going to press.

Published by Collins
An imprint of HarperCollins*Publishers* Ltd
1 London Bridge Street
London SE1 9GF

HarperCollins*Publishers* Macken House
39/40 Mayor Street Upper, Dublin 1,
DO1 C9W8, Ireland

© HarperCollins*Publishers* Limited 2021

ISBN 9780008326708

First published 2015

This edition published 2021

10 9 8 7 6 5 4

British Library Cataloguing in Publication Data.

A CIP record of this book is available from the British Library.

Publishers: Katie Sergeant and Clare Souza
Project Management: Richard Toms and Rebecca Skinner
Author: Trevor Senior
Cover Design: Sarah Duxbury and Kevin Robbins
Inside Concept Design: Sarah Duxbury and Paul Oates
Text Design and Layout: Jouve India Private Limited
Production: Karen Nulty
Printed in the United Kingdom by Ashford Colour Press Ltd

MIX
Paper | Supporting responsible forestry
FSC™ C007454

This book contains FSC™ certified paper and other controlled sources to ensure responsible forest management.

For more information visit: www.harpercollins.co.uk/green